# United Kingdom Climate Change Impacts Review Group

**M.L. Parry**
*Chairman*
University of Birmingham

**P. Bullock**
Soil Survey and Land Research Centre

**M.G.R. Cannell**
Institute of Terrestrial Ecology (Natural Environment Research Council)

**A.F. Dlugolecki**
General Accident Fire and Life Assurance Corporation p.l.c.

**G. Marshall**
Dobson Park Industries p.l.c.

**J. Page**
University of Cambridge

**A. Perry**
University College Swansea

**D. Potts**
University of Nottingham

**D. Pugh**
Institute of Oceanographic Science

**J.F. Skea**
University of Sussex

**K. Smith**
University of Stirling

**P. Turton**
Water Services Association

**M.H. Unsworth**
University of Nottingham

**R. Warrick**
University of East Anglia

**S.M. Cayless**
*Technical Secretary*
Department of the Environment

**R.B. Wilson**
*Executive Secretary*
Department of the Environment

The authors accept liability for the content of this report but the views expressed are their own and not necessarily those of the organisations to which they belong or the Department of the Environment.

# The Potential Effects of Climate Change in the United Kingdom

**United Kingdom**
**Climate Change Impacts Review Group**

**First Report**

Prepared at the request of the
Department of the Environment
January 1991

LONDON: HMSO

# Contents

# Executive Summary

## BACKGROUND

1. The Climate Change Impacts Review Group (CCIRG) was set up to consider the potential impacts of climate change in the United Kingdom, and to establish what further research is needed.

2. The report considers the potential impacts of climate change in a wide variety of environmental and socio-economic areas.

3. Following the general approach adopted by the Intergovernmental Panel on Climate Change (IPCC) the Review Group considered those impacts estimated to occur under a change of climate assuming continued present global rates of increase in greenhouse gas emissions (a 'Business-as-Usual' scenario). Policies to reduce emissions, and the environmental and socio-economic effects of these policies, are not considered in this first report.

4. There are major uncertainties concerning possible greenhouse gas – induced climate change and its associated impacts. This report should be considered to be a preliminary assessment. As more information becomes available the assessment of potential impacts will inevitably be revised and improved.

## CHANGES IN CLIMATE AND SEA LEVEL

1. Scenarios of UK climate change in the years 2010, 2030 and 2050 are considered. These scenarios are **not** predictions but rather possible outcomes suggested by 'state-of-the-art' climate models. This approach, and the climate change scenarios reported here, are similar to those of the Intergovernmental Panel on Climate Change (IPCC). A 'Business-As-Usual' scenario of emissions of greenhouse gases results in an equivalent doubling of the pre-industrial $CO_2$ level in 2028. This implies the following:

- The mean global surface air temperature in about 2030 is estimated to be 0.7°C – 2.0°C higher than at present, with a best estimate of 1.4°C (2.2°C, in 2050).

- The atmospheric content of carbon dioxide ($CO_2$), important for its direct effect on vegetation, is estimated to increase from a present level of 350 parts per million by volume (ppmv) to reach about 450ppmv by 2030.

- In the summer season in the UK, the climate models indicate that temperature changes should be comparable to the global mean and spatially uniform over the UK and most of Europe.

- In the winter season the UK will experience enhanced warming with increasing temperature change along a southwest – northeast gradient. By the year 2030, winters in the UK could be approximately 1.5° – 2.1°C warmer than at present (2.3°C – 3.5°C in 2050).

- Extreme warm years could occur more often. For example, the occurrence of a hot summer similar to that of 1976 (the hottest summer on record in Central England) could increase one hundredfold by 2030 to a frequency of once in every 10 years.

- Precipitation in the UK is most likely to increase in winter. The best estimate is that, on average, winter precipitation will be 5% higher in 2030 (8% in 2050).

- As regards changes in summer precipitation, model results differ in direction as well as amounts. The best estimate is probably no change, but with a range of uncertainty of ± 11% in 2030 (± 16% in 2050). Even so, it is likely there would be greater evaporation in the warmer climate and thus drier soils.

- The best estimate is that global mean sea level will be about 20 cm higher than today by the year 2030 (30 cm by 2050). This estimate is broadly applicable to the UK but must be adjusted for ongoing vertical land movements in specific coastal locations.

2. The Review Group emphasises that the uncertainties in model-based predictions of regional climate changes are large. For instance, the seasonal and spatial details of current precipitation patterns are generally not well simulated, and the predictions of greenhouse gas-induced changes in precipitation vary greatly from one model to another. Some aspects of climate, such as regional storms, rainfall intensity and winds, involve processes which occur over relatively small areas and are not well simulated within the coarse resolution of current climate models. Greater confidence can be attached to changes in climate means than to changes in climate variability (including extremes).

3. Even if stringent measures are taken to reduce greenhouse gas emissions, some changes in climate and sea level will occur anyway – a 'climate change commitment'. This is due to time-lags in the climate system response to changes in greenhouse gas concentrations that have already occurred.

## OVERALL CONCLUSIONS

1. The Review Group concludes that changes in climate could lead to significant impacts in some sectors and environments and in certain regions of the UK. There will be beneficial as well as adverse effects. However, the Review Group made special effort to identify those aspects of climate change which pose particular risks. More specifically the Review Group concluded the following:

2. *Climatically,* most critical are the potential changes in extreme conditions such as hot spells, droughts and storms. Given the scenarios of UK climate change noted above, it is likely that the occurrence of hot, dry summers would increase, while the chances of extreme cold winters would decrease. For other extremes, such as severe storms, the scientific 'state-of-the-art' precludes even reasoned speculation. In developing climate change predictions high priority should be given to changes in extremes.

3. *Environmentally,* the potential effects of climate change on hydrology – water quantity, water quality, evapotranspiration and soil moisture – are of critical importance, with serious implications for nearly all sectors in the UK. Soils, flora, fauna and associated natural ecosystems are generally very sensitive to variations in moisture. Anticipating changes in the availability of water as a result of changing climate is important for long-term planning in all sectors but is crucial for agriculture, horticulture, forestry and the water industry.

4. *Geographically,* the coastal zone deserves special attention as an area potentially at risk. Coastal and estuarine areas will be affected by sea level rise as well as climate change, and contain a large proportion of the UK population, manufacturing industry, energy production facilities, valued natural environments such as beaches and wetlands, recreational amenities and mineral extraction sites. There is a possibility that changes in climate and sea level would exacerbate existing problems such as coastal erosion, storm surge damage,

degradation of wetlands and estuaries, and groundwater contamination.

5. *Sectorally,* there are considerable differences in sensitivity to climate change, and sectors are not equally amenable to adaptation. There are three broad categories:

i) Sectors which are likely to be most seriously affected are: soils, water, natural flora and fauna. There could be potentially large changes in landscape and habitat with some species being favoured to the detriment of others.

ii) Sectors in which considerable changes are likely but where adaptation is clearly possible include: forestry, agriculture and coastal infrastructure.

iii) Sectors in which impacts may be smaller and where there may be some positive benefits as well as costs include: energy, manufacturing, construction, minerals extraction, transport, finance, recreation and tourism.

These conclusions have clear implications for setting priorities in the planning of adaptive strategies. Particular attention should be focused on the management of natural and semi-natural ecosystems and on those sectors where adaptation will prove most challenging.

## RECOMMENDATIONS

*A. Implications for Policy Development*

1. Although the potential impacts of climate change in the UK are only partly understood, the risks of commitment to these unknowns could be minimised at both national and international levels by (i) developing strategies to mitigate these impacts by abating greenhouse gas emissions, (ii) formulating appropriate adaptive strategies, and (iii) taking advantage of any beneficial impacts likely to arise. Sustainable development will require careful consideration of the balance between preventive and adaptive strategies, their timing and their cost.

2. Small increases in temperature could significantly alter the species composition in over half of the statutory protected areas in the UK (National Nature Reserves, Sites of Special Scientific Interest, etc.). Management policies aimed at preserving the

existing communities within reserves may need to be reviewed. Consideration needs to be given to the active transfer of species and to the creation of new reserves.

3. In the forestry and farming sectors, current policies concerning land use, water resources planning and forest development may need to be reviewed in the context of a changing climate. The appropriate information needs to be assembled to support such a review.

4. The construction or renewal of sea defences should take account of changes in flood probabilities in their lifetimes, without excessive additional expense.

5. Water resources (e.g. reservoirs) may need to be developed and managed to deal with altered rainfall patterns. This would have implications for water charges and demand management, and put pressure on land use development.

6. The potential impact of climate change on energy demand is substantial. In particular space heating demand could decline. This factor should be considered in any long range energy scenario exercises that are undertaken.

7. Provisions should be made to ensure that emerging information on climate change and its effects is transferred to local authorities. This information should be in a form that is both comprehensible and relevant to local needs in order to provide guidance on future land use and structure planning.

8. There is a need to review the procedures by which design codes and standards for transport, construction, manufacturing industry and minerals extraction are revised to take realistic account of the impacts of climate change on future risks. The experience of the insurance industry in assessing and covering risks would be an important input to such a review.

9. Present practices for the inspection and monitoring of buildings, transport and water structures are based on an understanding of current levels of risk. These practices may need to be reviewed in the light of projected climate change.

10. At present there are significant variations between EC countries in the fiscal treatment of insurance catastrophe reserves, and in mechanisms for coping with catastrophe payments. With the possibility of natural catastrophes occurring more often as a result of potential changes in climate, these differences should be reviewed.

11. The possible global impacts of climate change should be communicated to policy makers in tourism and incorporated into the political and commercial planning horizons within the industry.

12. Research on the potential impacts of climate change is at present constrained by the cost of obtaining access to existing meteorological, sea level and other environmental data. These data should be made more readily available for research, at nominal cost.

B. Recommendations for Research

1. In the absence of detailed predictions of future climate change much valuable research can and needs to be done on the sensitivity of different sectors to climate change. Our knowledge could be extended by adopting a number of approaches.

i) There is a need to identify threshold values for critical processes influenced by climate change beyond which severe problems may develop. It is important that both the magnitude and the rate at which threshold values are reached should be identified.

ii) There is a need to examine the adaptive capabilties that currently exist within various sectors in the UK, and the policies that could be developed to foster these capabilities.

iii) The impacts of climate change experienced abroad could have indirect, though important, economic effects in the UK. As the understanding of potential climate change impacts evolves, research should begin to address this issue.

iv) A better understanding of the possible impacts of climate change in the UK can be gained by analysing the effects of historical and recent weather extremes in the UK and by studying analogous data in other countries already experiencing the climate conditions that may occur in the UK in the future.

v) Climate impacts are essentially local. Integrated assessments are required at the appropriate regional scale in order to understand the cascade of impacts through biophysical systems to the economic and social sectors.

2. Studies are required to find the balance between the costs of the climate change impact under a 'Business-As-Usual' scenario and the costs of cumulative measures to reduce the climate change, i.e., the balance between adaptation and prevention.

3. Further knowledge of changes in aspects of climate is critical in order to improve our understanding of potential impacts, in particular rainfall variability, both temporal and spatial, and variables such as cloudiness, windiness and storminess.

4. There is a particular need to investigate the effects of increasing levels of atmospheric $CO_2$ and temperature on regional evapotranspiration, soil water balance and the performance of trees, grasses and other crops.

5. There is a requirement to investigate the effects of climate change on the sensitivity of land to degradation processes, such as erosion, and the effects of possible sea level rise on land drainage.

6. More knowledge is required concerning the effects of climate change on potential and actual plant and animal distributions (including insect populations). To do this requires the development of predictive models based on data from both observations in the field and in controlled environments. Particular attention should be given to the rates at which species can migrate or adapt in response to climate change, and to the indentification of species or communities that may be lost.

7. Many of the opportunities to adapt current practices in agriculture and forestry to changes in climate depend on knowing how different species and their pests and pathogens respond to different types of weather, to increased concentrations of $CO_2$, and to varying supplies of water and soil nutrients. Research is needed to identify the changes in species or cultivars that may be required or permitted following climate change, and on the concurrent response of pests, pathogens and weeds both to these changes and to changes of climate.

8. The scenarios of sea level rise imply increases of only 2-7 centimetres per decade over the next 60 years. During this period comprehensive monitoring systems should be set in place, both globally and nationally. In addition, a high priority should be given to improving the predictive capability linking global temperature increases and other aspects of climate change to sea level rise and flood probabilities. Account should be taken of the impacts of these changes on the design of long-term coastal facilities. Even at this early stage it would be useful to initiate a series of local case studies in the UK to identify the most significant potential impacts and possible responses.

9. The water industry requires more accurate predictions of rainfall over the annual cycle and of the variability of extreme events. Research should incorporate those factors which affect the industry, such as alterations in soil structure and agricultural practice, especially in relation to fertiliser and pesticide use. An economic assessment of impacts of drought on agriculture and industry is required.

10. Further research is needed to quantify the impacts of climate change on patterns of energy demand and, in particular, the possible increase in electricity use for air conditioning. Research is also needed to assess the implications of altered patterns of energy demand for investment in energy supply, taking into account possible changes in peak demand for electricity and the fossil fuels. Any such analysis should be conducted within a scenario framework which considers the many structural changes in the energy sector, such as new patterns of power station fuelling, which are likely to take place. In addition, research is required to evaluate the effects of an altered climate on the availability and cost of renewable energy sources and to assess the vulnerability of coastal and estuarine energy facilities to sea level rise.

11. Research in particular locations is needed to quantify the impacts of climate change and sea level rise on manufacturing plant, housing, services and infrastructure. Further detailed investigation is needed to identify industries particularly sensitive to changes in temperature, water supply and water table, and to variations in river discharge.

12. Investigations are required of the effect of climate and sea level changes at coastal sites where there are facilities and operations for extracting, processing and refining minerals. Studies should examine the availability and cost of minerals for improving sea defences and effects of changes in the water table on the minerals extraction industry.

13. More knowledge is required concerning the effects of increased soil salinity, altered water table levels and soil shrinkage on current and future construction. The construction industry will need to be responsive to possible changes in heating, cooling and daylighting costs that may result from climate change. Research into hot weather problems in buildings and the role of improved glazing systems could avoid excessive reliance on the expanded use of air conditioning systems as a response to increases in temperature.

14. Comprehensive studies of the effects of climate change on transport have not been carried out and an assessment is required of the likely climate-induced demand for transport if life styles, residential and migrational patterns change. Requirements for winter rail and highway maintenance could change, and cost-benefit analyses of existing and future investment are required.

15. For the purposes of insurance, links between the levels of meteorological event and financial damage should be more precisely quantified and information on the value of property at risk must be improved. Research is needed into predictions of future probability distributions of climatic variables to allow assessment of the severity of different losses.

16. Baseline studies of recreation climatology, supplemented by behavioural surveys, are required to improve the understanding of the sensitivity of tourism to climate in the UK. Studies of weather-related holiday decisions, activity patterns and visitor responses to varying weather conditions at key tourist sites will all require basic research.

## SPECIFIC CONCLUSIONS

### Soils

1. The amount and availability of water stored in the soil is critical in determining land use. The water holding capacity of soils is likely to decrease and soil moisture deficits increase with higher temperatures. These changes would have a major effect on the types of crops, trees or other vegetation that soils in a particular area can support. The whole pattern of land use in the UK may change as a result.

2. Changes in soil moisture content have implications for the workability of land. Cultivation may become more difficult under drier conditions and new types of machinery may be needed. The period in which the soil is easily cultivable may also change, with implications for sowing and other practices associated with crop production.

3. Given drier, warmer summers, many soils would shrink more extensively than hitherto and, in winter under wetter conditions, would swell again. This alternate shrinkage and swelling would have important implications for the stability of foundations of buildings and other structures. The areas most affected would be central, eastern and southern England, where there are clay soils with a large shrink-swell potential.

4. Soil shrinkage under drier, warmer summers would lead to the formation of cracks linking the soil surface to underlying parent material. Such cracks would become the main pathways for movement of water, solutes and pollutants from the soil surface to depth with major effects on soil moisture balance, efficiency of fertilizers, herbicides and pesticides, and the rates at which water or pollutants reach watercourses.

5. Some soils will change rapidly. The rate of loss of organic soils would increase significantly, affecting ecological habitats nationally and agricultural use in southern England. Currently poorly drained mineral soils may become drier and hence less of a problem whereas, close to low-lying coastal areas, well-drained soils may become poorly drained.

6. Loss of organic matter and a decrease of soil biological diversity would reduce the stability of soil structure, resulting in increased vulnerability to land degradation.

7. If water tables rose in response to a rise in sea level, a change in soil processes would result. Existing land drainage schemes may become ineffective and there would be a need for new drainage schemes if land is to remain suitable for arable cropping.

8. Soils affected by rising sea levels would become saline and this would affect the types of crop that can be grown, as well as land management and the vulnerability of subsoil components of engineering structures to corrosion.

9. Although outside the remit of this Review, it should be emphasised that soils represent a significant source and sink for greenhouse gases. Climate change is likely to alter the balance between the source and sink roles of soils in such a way as to lead to a significant increase in the release of these gases.

### Flora, Fauna and Landscape

1. The natural biota are sensitive to changes in almost all aspects of climate and weather. The parameters to which they would be most sensitive are: extreme weather events or seasons, soil water deficits, changes in mean temperature and increases in atmospheric $CO_2$ concentration. Any sustained rise in mean temperature exceeding 1°C will have very significant effects on the UK flora and fauna.

2. Where rainfall is plentiful, an increase in $CO_2$ concentration accompanied by a lengthening of the growing season is likely to increase the productivity of vegetation. High $CO_2$ levels would also alter plant development and water use in ways that are not yet understood.

3. There may be significant movement of species northwards and to higher elevations. The rate of northward movement for the climatic scenarios given is potentially 100 km/decade although few species are capable of migrating at this rate. Insects (invertebrates), birds and 'weedy' plant species will move first. Species which do not extend their ranges will not necessarily be affected adversely – many species will adapt to climate change. However, the predicted rate of climate change is too rapid for many species (e.g. most trees) to adapt genetically.

4. There will be an increased probability of invasion and spread of alien weeds, pests, diseases and viruses, some of which could be potentially harmful.

5. Many native species and communities will be adversely affected and may be lost to the UK. The losses (of both plants and animals) will occur particularly within (i) montane communities, (ii) salt marsh and coastal communities, (iii) confined 'island' habitats, and (iv) wetlands and peatlands.

6. An increase in sea level of 20-30 cm would have an impact on about 10% of notified natural reserves, and might adversely affect invertebrates and birds which inhabit mudflats.

7. There may be an increase in the overall number of species of invertebrates, birds and mammals in much of the UK (following migration and invasion), but the natural flora is likely to be impoverished by the loss of many endangered species which occur in isolated damp, coastal or cool habitats.

8. Land-use changes brought about by changes in agricultural and forestry policy in the UK could influence the natural biota as much as climate change itself.

*Agriculture, Horticulture, Aquaculture and Forestry*

1. Experiments in controlled environments suggest that an increase of atmospheric $CO_2$ from the current value of 350 ppmv to about 450 ppmv by 2030 could increase the productivity of crops by 5-15%. However, it is uncertain whether this increase will occur in the field.

2. With increased $CO_2$ levels, the amount of water used per unit of plant growth decreases but, because plants growing in high $CO_2$ are larger than those in ambient air, it is not clear whether the *total* water requirement of crops and forests would decrease.

3. In general, even if there was sufficient soil water, higher temperatures would decrease the yields of cereal crops (such as wheat) although the yield of crops such as potatoes, sugar beet and forest trees would tend to increase. The introduction of new cultivars better suited to the changed climate could optimise these responses.

4. The length of the growing season for grasses and trees would increase by about 15 days per degree C increase in mean temperature. This increase could improve the viability of grassland, animal production and forestry in the uplands. Higher temperatures would also create opportunities to convert currently unproductive upland moor and heath to grass pasture, but this would require considerable inputs of lime and fertilizer.

5. More than 20% (by value) of UK horticultural crops are grown unprotected by glasshouses. A warmer climate would offer opportunities for raising high value overwintered crops and for shortening the growth period of spring-sown crops. Glasshouse heating costs would be reduced, but any increase in winter cloudiness could decrease the production potential in glasshouses.

6. Problems such as colorado beetle on potatoes and rhizomania on sugar beet, currently thought to be limited by temperature, could become more important in the future, as could the risk of pest outbreaks on forests.

7. If temperatures increase, there would be opportunities to introduce into the UK new tree species and crops that are currently grown in warmer climates. Maize and sunflower might be grown for their seed/grain yield as well as for fodder over much of the arable area of the UK if temperatures rose by about 1.0° – 1.5°C from present values. Workability of land and weather extremes (wind, frost) might constrain their introduction.

8. Some forests or woodlands growing on poor, dry soils may become unhealthy or die following a succession of dry summers, despite longer growing seasons and $CO_2$ fertilization.

9. Changes in climate may have large effects on soils. Land use policy may need to be reviewed in the light of this, with regard to the changing potential of soils for crops, trees and other plant species. Guidelines for decisions on the protection of certain soils and land may need to be developed. Monitoring of land and soil salinity should be incorporated into soil conservation policy.

10. An increase in the frequency of long hot summers would be detrimental to the trout and salmon farming industries, both directly by limiting water availability, and indirectly through increased incidences of low oxygen concentrations in water and of disease.

*Coastal Regions*

1. Increases in mean sea level would lead to an increased frequency of extremely high sea levels and coastal flooding. If there were also increases in storminess, storm surges and waves would further increase flooding probabilities.

2. Flooding would result in damage to structures and short-term disruption of transport, manufacturing and the domestic sector. In addition, longer term damage to agricultural land, engineering structures such as buildings and coastal power stations, rail and road systems, would occur in some areas due to saline effects.

3. A number of low-lying areas are particularly vulnerable to sea level rise: these include, for example, the coasts of East Anglia, Lancashire and the Yorkshire/Lincolnshire area, the Essex mudflats, the Sussex coastal towns, the Thames estuary, parts of the North Wales coast, the Clyde/Forth estuaries and Belfast Lough.

4. An increase in sea level would be likely to reduce the efficiency of groundwater and sewage drainage in low-lying areas, with consequent need for more pumping.

5. Groundwater provides about 30% of the water supply in the UK. A rise in the water table associated with higher sea levels may increase the salination of groundwater over a long period, with consequent effects on agriculture and water supply.

6. There is the possibility of the inundation of significant areas of wetlands and saltmarshes found near the present High Water line, with the probability of concomitant inshore migration of coastal ecosystems.

7. The success of adaptive responses to changes in coastal flooding frequencies will depend on the rate at which these changes develop. This applies both for natural systems such as salt marshes, and for socioeconomic activities such as manufacturing, mining, transport and recreation.

*Water Industry*

1. Wetter winters would benefit water resources in general, but warmer summers with longer growing seasons and increased evaporation would lead to greater pressures on water resources, especially in the south and east of the UK. Increased variability in rainfall, even in a slightly wetter climate, could lead to more droughts in any region of the UK.

2. At present water demand in warm months may be 25% above average and, locally, a doubling of demand can occur on a hot dry day. Higher temperatures would lead to increased demand for water and higher peak demands, requiring increased investment in water resources and infrastructure if restrictions were to be avoided.

3. An increase in temperature would increase the demand for irrigation. In times of drought the abstraction for agriculture competes with abstractions for piped water supply by other users.

4. A large range of industries use water in production. Some industrial activity (e.g. chemicals production and food processing) would be severely curtailed if water supply were interrupted as a result of drought.

5. Increases in the frequency of drought could have an impact on public health through interruptions to domestic and other supplies.

6. Climate change may lead to changes in soil structure causing increased leaching, decreased absorption, and changes in agricultural applications of fertiliser and pesticides which could affect river, lake and ground water quality, possibly adversely.

7. Many existing power stations rely on river water for cooling. Climate change may reduce the availability of cooling water, particularly during summer months, with possible implications for the operation of individual power stations.

*Energy*

1. The lifetimes of many energy-producing or consuming items of equipment are shorter than the

timescales over which climate change might occur. With adequate preparation, successful adaptation to changed conditions is likely.

2. Higher temperatures would have a pronounced effect on energy demand. Space heating needs would decrease substantially. Given the present pattern of heating energy supply, natural gas use would decrease most.

3. Higher demand for air conditioning would entail greater electricity use. In the longer term the net effect of changed space heating and air conditioning needs could be an increase in electricity demand.

4. Changes in building design and technological developments in heating and air conditioning equipment stimulated by climate change could have secondary impacts on energy demand.

5. Total UK energy demand is likely to fall as a result of climate change. The effect on energy prices is not known, but is likely to be small. The consequences for different classes of consumer depend on: patterns of use changes in the energy supply system, the rate of depletion of natural resources, and changes in global energy markets.

6. The vulnerability of coastal or estuarine energy facilities to storm surges is likely to be increased by sea level rise. All UK petroleum refineries and half of the present power station capacity are located on the coast or on estuaries. There may be a need to strengthen sea defences at certain locations in order to protect against flooding.

*Minerals Extraction*

1. Projected changes in temperature and rainfall are unlikely to result in significant technical problems for the oil and gas, coal, industrial minerals and aggregates industries in the UK. Minerals are extracted successfully throughout the world under a wide range of climatic conditions and appropriate working practices and technology could be adapted to the UK. In addition, UK engineers and management would have adequate time to develop any new techniques that may be required to accommodate the likely impacts.

2. Rises in sea and estuary levels and changes in the water table and salinity would, in some areas, be very important to land-based mineral extraction and dredging. There could be increases in pumping and mineral processing costs, and the restoration of extraction sites would be more complicated.

3. The projected rates of rise in sea and estuary levels are sufficiently low to allow investments in current operations to be recovered provided that there is no increase in the intensity and frequency of storm surges.

4. Decisions to improve sea and river defences would require increased UK production, and possibly the increased import, of larger sizes of hard stone. For example, recent damage to sea defences at Towyn, in North Wales, gives an indication of the impact of storm surges. Boulders weighing up to five tonnes were required for repair and these are larger than usually quarried.

5. Changes in the energy market in connection with space heating and air conditioning requirements due to temperature changes would affect the demand for coal, oil and natural gas and would imply changes in the structure of the UK minerals extraction industry.

*Manufacturing*

1. Manufacturing industry contributes approximately 30% of the UK's Gross Domestic Product and accounts for over 35% of the UK's income from exports. It is therefore an important sector to consider in relation to the impacts of climate change.

2. Most of manufacturing should be able to adapt to climate change. Climate change will produce conditions in the UK which already exist elsewhere in Europe where manufacturing industry contributes effectively to GDP and competes internationally.

3. The major concern for manufacturing industry, 40% of which is located in coastal and estuarine areas, relates to the potential for sea level rise. For rates of rise up to about 5cm per decade the costs could probably be absorbed, but rates in excess of this and over the longer term could substantially increase costs. These impacts will differ greatly from site to site.

4. Increases in temperature, with reduced winter freeze, would be likely to improve productivity in some areas such as the building industry, to improve the transport of raw materials and finished products, and to decrease disruptions in production due to weather. However, the food industry could be affected by locational changes in produce supply and may find an increased need for refrigeration.

5. Adverse effects may arise in certain manufacturing industries due to shortage of water (e.g.

paper making, brewing, food industries and power generation). Repeated annual droughts could adversely affect their production.

## Construction

1. Warmer winters would reduce heating energy demands in buildings by significant amounts (about 30–40%).

2. Warmer winters would confer some benefits for construction productivity, as the industry would become less affected by snow and ice.

3. Increased winter rainfall would adversely affect building operations.

4. Unless buildings are better designed to take account of warmer summers, interiors could become significantly warmer in summer. This is likely to increase the demand for air conditioning and the consequent use of electrical energy.

5. Buildings are vulnerable to extreme climatic events such as high winds and severe wind-driven rain, but it is currently not possible to assess changes in long term risks due to climatic extremes other than temperature extremes. However, some risks are likely to decrease, for example, snow loading.

6. The greatest negative impacts on construction are likely to arise from the combination of sea level rise interacting with alterations in inland water hydrology. The impacts will be localised in vulnerable areas.

7. Foundation stability on shrinkable soils would be affected by increased winter rainfall combined with drier summer soil conditions. South eastern England will remain the area with most properties at risk.

8. Design risk codes, currently based on historical climatic experience, will in future need to match the assessd risks emerging from the impacts of climate change.

## Transport

1. Sensitivity to weather and climate change is high for all forms of transport, but especially for road and air transport.

2. A reduction in the frequency, severity and duration of winter freeze would reduce disruption to transport systems. Winter maintenance expenditure on UK roads should decrease, with less salting and snow-clearing required. A decrease in the number of freeze-thaw cycles would result in reduced road damage.

3. Snow and ice present the most difficult weather-related problems for the railway system: these problems could be expected to be reduced, with potential savings in locomotive rolling stock design, point heaters and de-icing equipment. Increases in temperature would also reduce snow and ice problems that hinder aircraft operations, although changes in prevailing winds could affect runway operations or re-distribute noise impacts in urban areas.

4. Any increase in the frequency of severe gale episodes could increase disruption from fallen trees, masonry and overturned vehicles, and interrupt flight schedules and airport operations.

5. Sea level rise and more frequent coastal flooding could cause structural damage to roads, bridges, embankments and other transport infrastructure.

6. If precipitation increased, this could exacerbate road flooding, landslips and corrosion of steelwork on bridges.

7. Changes in the demand for some goods, e.g. perishable foodstuffs, coupled with higher ambient temperatures, may affect the pattern and frequency of distribution of goods to wholesalers and retailers.

8. Changes in demand for travel, such as increased leisure journeys, may result from perceived 'better' summer weather.

## Financial Sector

1. There is little information on the potential impacts of climate change upon the various activities within the financial sector, except for the insurance industry, which by the nature of its business would be immediately affected by a shift in the risk of damaging weather events arising from climate change. The cost of severe weather events to insurers and reinsurers has risen steeply (over £1 billion from the October, 1987 'hurricane' alone), with the domestic sector presenting the greatest losses.

2. Because of the international nature of financial markets and institutions, the impacts from climate

change or sea level rise abroad are likely to be just as important (perhaps even more important) than the impacts in the UK itself.

3. If the risk of flooding increases due to sea level rise, this would expose the financial sector to the greatest potential losses (the value of the property protected by the Thames Barrier is £10 to £20 billion).

4. If the probability distribution of individual variables such as temperature, rainfall and windspeed changes, then this would alter the frequency of severe weather damage events, and may necessitate pricing or product changes.

5. If there were an increased frequency of multiple incidents (e.g. windstorms plus sea surges in the UK or internationally within successive years), this would place financial strain on the insurance industry. The persistent storminess during January and February 1990 led to record losses.

6. There will be increasing pressure to improve the quality of information about property exposed to damage. In turn, information on actual weather damage impact will be useful for property owners and design engineers, in order to mitigate or avoid future losses.

7. If climate change and sea level rise seem likely to affect socioeconomic activities, this will affect the appraisal of existing and new investment opportunities, particularly in agriculture and for coastal regions, and it could also alter the role of insurance in protecting the most vulnerable areas or activities.

8. Financial operations are vulnerable to short-term disruption due to failure of communications or denial of physical access, such as resulted in the October 1987 storm in the London area.

*Recreation and Tourism*

1. Tourism in the UK has an international dimension which is sensitive to any change in climate that alters the competitive balance of holiday destinations worldwide.

2. Increases in temperatures are likely to stimulate an overall increase in tourism in the UK, with the greatest effects on activity holidays and some forms of outdoor recreation.

3. If any change towards warmer, drier summer conditions occurs there may need to be more restricted public access to large areas of upland Britain because of the enhanced risk of fire.

4. Increases in temperature in the UK will have an effect in upland Britain, where the balance between outdoor recreation and agriculture may be altered, if, for example, more intensive agriculture reduces public access to the countryside.

5. The commercial viability of some winter sport developments in upland Scotland would be at risk if snowfall and snowcover are reduced.

6. Any significant increases in rainfall, windspeed or cloud cover could offset some of the general advantages expected from higher temperatures.

7. Rises in sea level would affect fixed waterfront facilities such as marinas and piers. For beaches backed by sea walls, increased erosion would lead to a lowering of the beach and subsequent undermining of the walls. Recreational habitats such as sand dunes, shingle banks, marshlands and soft earth cliffs would also be affected.

8. An increase in sea temperatures would increase the pressures of tourism on UK beaches, while coastal erosion may reduce beach area.

# Introduction

## 1.1 BACKGROUND

A greater emphasis has recently been placed on the quality of the natural and man-made environment and the need to pass to future generations that same quality of environmental 'capital' that the current generation inherited. This concept of sustainable development was discussed in the Pearce Report to the Department of the Environment (Pearce, *et al.*, 1989). One of the most important areas identified in the Pearce Report as needing urgent attention was the possible impact of climate change. The Climate Change Impacts Review Group (CCIRG) was established to inform Ministers of HM Government of the potential impacts of climate change in the United Kingdom, to identify research that is needed and the issues of policy that should be addressed.

This first report of CCIRG gives an indication of the likely impacts of climate changes on the environment and economy, suggests measures that may be needed to prevent, mitigate or counteract these changes, and identifies gaps in knowledge where more research is required. It also attempts to assign some sense of priority in terms of economic impact and timescales. The report builds on previously commissioned reviews and workshops, including meetings in Birmingham in 1987, Wallingford in 1988, and Darlington in 1989 (Parry and Read 1988; Department of the Environment, 1988).

## 1.2 SCOPE OF THE REPORT

The Review Group was set up as an independent body of experts (Annex 3). The members of the Group were able to call on the expertise of a number of experienced individuals (Annex 4).

The remit of the Review Group was to:
(i)  Assess the impacts that would be likely to occur without significant policy responses, under changes of climate that are currently estimated to occur unless there are major changes in greenhouse-gas emissions.

(ii)  Identify critical uncertainties and unknowns that currently limit our ability to estimate likely impacts or respond to them.

iii)  Consider the implications for policy.

iv)  Make recommendations for future research.

The areas covered include: changes in climate and sea level; soils; flora, fauna and landscape; agriculture, horticulture, aquaculture and forestry; coastal regions; water; energy; minerals extraction; construction; transport; the financial sector; recreation and tourism. This report does not consider the health implications of climate change, which is the subject of another review.

The report estimates impacts under a 'Business-As-Usual' increase in greenhouse-gas emissions which is compatible with that adopted by the Intergovernmental Panel on Climate Change (IPCC). This approach assumes that current emission growth rates remain at their present levels over the period of projection with the exception of CFCs on which action is already being taken. For the latter, it was assumed that the Montreal Protocol would apply and that alternatives to CFCs would be adopted. With this approach, an equivalent doubling of the pre-industrial $CO_2$ level is estimated to occur in about 2030.

Because of the time constraints applied, the report does not examine control strategies and adaptive measures, or the interrelationships between the possible international impacts of climate change and the national impacts discussed herein. Nor does it consider the potentially complex interactive effects of climate change and other, ongoing, environmental trends, such as acidification and groundwater pollution. These are important subjects for future study.

# Future Changes in Climate and Sea Level

# 2

**SUMMARY**

- Scenarios of UK climate change in the years 2010, 2030 and 2050 are considered. These scenarios are **not** predictions but rather possible outcomes suggested by 'state-of-the-art' climate models. This approach, and the climate change scenarios reported here, are similar to those of the Intergovernmental Panel on Climate Change (IPCC). A 'Business-As-Usual' scenario of emissions of greenhouse gases results in an equivalent doubling of the pre-industrial $CO_2$ level in 2028. This implies the following:

  - The mean global surface air temperature in about 2030 is estimated to be 0.7°C–2.0°C higher than at present, with a best estimate of 1.4°C (2.2°C, in 2050).

  - The atmospheric content of carbon dioxide ($CO_2$), important for its direct effect on vegetation, is estimated to increase from a present level of 350 parts per million by volume (ppmv) to reach about 450ppmv by 2030.

  - In the summer season in the UK, the climate models indicate that temperature changes should be comparable to the global mean and spatially uniform over the UK and most of Europe.

  - In the winter season the UK will experience enhanced warming with increasing temperature change along a southwest–northwest gradient. By the year 2030, winters in the UK could be approximately 1.5°–2.1°C warmer than at present (2.3°C–3.5°C in 2050).

  - Extreme warm years could occur more often. For example, the occurrence of a hot summer similar to that of 1976 (the hottest summer on record in Central England) could increase one hundredfold by 2030 to a frequency of once in every 10 years.

  - Precipitation in the UK is most likely to increase in winter. The best estimate is that, on average, winter precipitation will be 5% higher in 2030 (8% in 2050).

  - As regards changes in summer precipitation, model results differ in direction as well as amounts. The best estimate is probably no change, but with a range of uncertainty of ±11% in 2030 (±16% in 2050). Even so, it is likely there would be greater evaporation in the warmer climate and thus drier soils.

  - The best estimate is that global mean sea level will be about 20 cm higher than today by the year 2030 (30 cm by 2050). This estimate is broadly applicable to the UK but must be adjusted for ongoing vertical land movements in specific coastal locations.

  - The uncertainties in model-based predictions of regional climate changes are large. For instance, the seasonal and spatial details of current precipitation patterns are generally not well simulated, and the predictions of greenhouse gas-induced changes in precipitation vary greatly from one model to another. Some aspects of climate, such as regional storms, rainfall intensity and winds, involve processes which occur over relatively small areas and are not well simulated within the coarse resolution of current climate models. Greater confidence can be attached to changes in climate means than to changes in climate variability (including extremes).

  - Even if stringent measures are taken to reduce greenhouse gas emissions, some changes in climate and sea level will occur anyway – a 'climate change commitment'. This is due to time-lags in the climate system response to changes in greenhouse gas concentrations that have already occurred.

## 2.1 INTRODUCTION

Due largely to human activities, the atmospheric concentrations of carbon dioxide, methane, nitrous oxide, and the chlorofluorocarbons have been increasing. As a result of deforestation and, particularly, the burning of fossil fuels, the concentration of carbon dioxide now stands at about 354ppmv (1990), a 25% increase over the pre-industrial (mid-18th century) level. The methane concentration has more than doubled over the same period. These so-called 'greenhouse gases' (GHG) are radiatively active and change the heat budget of the atmosphere, with the result that, on average, the Earth's surface and lower atmosphere should warm as atmospheric concentrations increase. The purpose of this Section is to address the issue of what global warming may mean for the climate of the United Kingdom.

## 2.2 ESTIMATING REGIONAL CLIMATE CHANGE: PROBLEMS AND APPROACH

Although it is certain that changes in climate will not occur uniformly around the world, there are large uncertainties involved in predicting the regional patterns of change (Mitchell *et al.,* 1990; Dickinson, 1986). The principal tools used to examine the sensitivity of climate to greenhouse gases, general circulation models (GCMs), do provide global coverage of regional climate changes. Unfortunately, different GCMs often give very different results. (Even if they all gave similar results, in the absence of careful model validations it would not necessarily mean that they were all correct results). Moreover, GCMs encounter major difficulties in correctly simulating the current climate at the regional scale, which leaves doubt about their predictions of future regional climate change.

Furthermore, sensitivity experiments with CGMs usually entail an instantaneous doubling of $CO_2$ and running the model until little additional change takes place. This provides information on the *equilibrium* global temperature response ($\triangle T_{2x}$) of the climate system to a $CO_2$ doubling, but says nothing about the rate of warming. This is unlike the real world in which both greenhouse gas concentration changes and climate changes are time-dependent. In estimating time-dependent changes in climate, the role of the oceans is particularly important. The oceans have a large capacity to transport and absorb heat, and as the world warms some of the heat is mixed into the deeper ocean layers. This creates a lag in the climate system, slowing the rate of global warming.

In order to estimate the actual time-dependent, or *transient,* global warming ($\triangle T$), the ideal procedure would be to run a coupled, dynamic ocean-atmosphere GCM with gradual increases in greenhouse forcing. A few such model experiments have now been performed (e.g. Washington and Meehl, 1989; Stouffer *et al.,* 1989). Although the magnitudes of the climate changes at the time of $CO_2$ doubling are less due to the oceanic lag effect, the spatial patterns generally resemble those of the equilibrium results obtained with the GCM. The exception is around Antarctica and the northern North Atlantic where the warming is reduced, but this has little effect on the predicted changes for the UK (Bretherton *et al.,* 1990). The major drawback to the use of ocean-atmosphere GCMs for deriving time-dependent climate change scenarios is the enormous amount of computing time required.

Instead, for this purpose, relatively simple upwelling-diffusion energy-balance climate models are most often used (Hoffert and Flannery, 1985). These models are forced by time-varying changes in greenhouse gas concentrations, and many of the processes involved in the mixing of heat into the deeper layers of the oceans are represented as a diffusive process. In one comparison, the predicted changes in global-mean temperature from such a model were shown to be in excellent agreement with those obtained from a more complex ocean-atmosphere GCM run in transient mode (Bretherton *et al.,* 1990). But while providing estimates of the rate of global warming, the simpler climate models cannot provide any regional details of climate change.

In order to provide scenarios of climate change for the UK, the time-dependent and equilibrium model results described above were combined. First, the averaged spatial patterns of five equilibrium GCM results, interpolated to a common $5° \times 5°$ grid and standardised with respect to their $\triangle T_{2x}$ values (to give the regional change in climate per degree warming) were obtained (from Santer *et al.,* 1990); Wigley *et al.,* 1990) (see Footnote 1). This average spatial pattern was then scaled by the time-dependent global-mean $\triangle T$ obtained from a simpler climate model. The big assumptions in this procedure are: that the average of five GCM results is a better representation of reality than any single model result, and that under continuous greenhouse forcing changes, the spatial patterns will not vary through time 'on the way' to the equilibrium condition – an unlikely situation, but the best that can be assumed at present.

It is necessary to emphasise that this procedure produces *scenarios* (what would occur), *not* predictions (what will occur).

4

## 2.3 GLOBAL WARMING PROJECTIONS

As a starting point for scenario development, projections of future changes of global-mean temperature are required. These are shown in Figure 2.1.

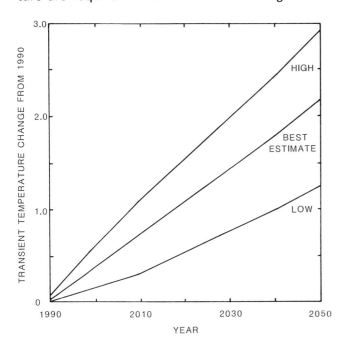

Figure 2.1 Projected global mean temperature change, 1990–2050. Based on Business-As-Usual greenhouse forcing, using climate sensitivity values of 2, 3 and 4 degrees (for the low, best estimate and high projections respectively).

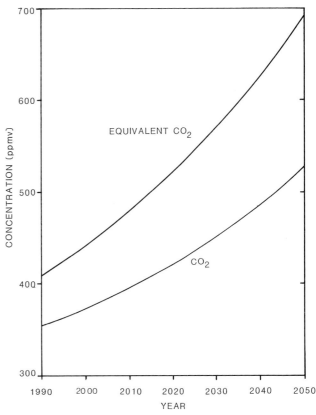

Figure 2.2 The Business-As-Usual greenhouse gas forcing scenarios, 1990–2050. The lower curve shows the projection of atmospheric $CO_2$ concentration (ppmv), assuming a continuation of the present fossil fuel emission growth rate of 1.5% with a constant biospheric emission of 1.5GtC. The top curve shows the combined radiative forcing changes from all the greenhouse gases, in equivalent $CO_2$ concentration.

The projected $\triangle T$ values were derived using the climate model of Wigley and Raper (1987). This is the same model used by the Intergovernmental Panel on Climate Change (IPCC) (Houghton *et al.,* 1990) (note: the IPCC obtained slightly different $\triangle T$ values due to differences in assumptions and forcing scenarios, see Footnote 2).

To force the model, a 'Business-As-Usual' (B-A-U) scenario of increasing greenhouse gas emissions was used. For this scenario, it was assumed that current emission growth rates remain in effect over the entire projection period. For example, the annual growth rate of fossil fuel emissions was held constant at 1.5%, with an unchanged rate of net biospheric emissions from deforestation of 1.5 gigatonnes of carbon (GtC) per year (the resultant $CO_2$ concentrations are shown in Figure 2.2). The exception to this assumption relates to the CFCs, for which it was assumed that the 1987 Montreal Protocol would apply and non-greenhouse-gas alternatives would be adopted. With this B-A-U scenario, an 'equivalent' doubling (taking into effect all greenhouse gases) of the pre-industrial $CO_2$ level occurs in about the year 2028 (Fig 2.2).

In the simpler climate models, a key 'tunable' parameter is $\triangle T_{2x}$, the 'climate sensitivity'. Different values of $\triangle T_{2x}$ reflect differences in assumptions regarding climate 'feedbacks' — changes in water vapour, albedo, clouds. It is generally agreed that most of the feedbacks are positive and, on the whole, enhance the initial, direct greenhouse-gas-induced warming. To reflect the scientific uncertainty in this value, a range of 2–4°C was used, with a best estimate of 3°C.

With these assumptions, the best estimate with the B-A-U scenario is that the world will be, on average, 0.7°C, 1.4°C and 2.1°C warmer than today in years 2010, 2030 and 2050 respectively.

## 2.4 UK WARMING SCENARIOS

As discussed above, the time-dependent projections of global warming are used to scale the

5

standardized spatial patterns of equilibrium temperature change as derived from the average of the five GCMs. The resultant patterns of winter and summer temperature changes are shown in Figure 2.3.

comparison, a number of individual extreme years are noted. For example, the hot summer of 1976 (17.8°C) was extremely rare in the central England temperature record, occurring only about once in a thousand years on average. The rarity of the cold

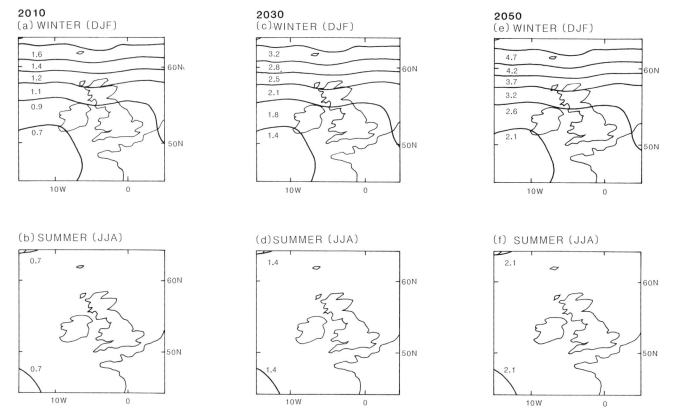

Figure 2.3 Seasonal mean temperatures for the UK for 2010, 2030 and 2050 (for explanation, see text).

## Summer

For the summer season (June, July, August), the average temperature change over the entire UK is comparable to the global-mean value. The absence of contours in the body of Figures 2.3 b, d, & f is an indication of the flatness of the temperature pattern. It means that over the whole region the $\triangle T$ is approximately 0.7°C, 1.4°C and 2.1°C for the years 2010, 2030 and 2050 respectively. This spatially uniform pattern of change implies that the current differences in summer temperature between, say, Scotland and southern England, will persist. By the year 2030, the mean summer temperature in central England is projected to be 16.7°C, which is equivalent to the current summer temperature of Stuttgart.

For assessing impacts, it is useful to know something about the frequency distributions of temperatures and how they will change. The distribution of mean summer temperatures is shown in Figure 2.4a, based on the 1660–1989 temperature record for central England (Manley, 1974; Jones, 1987). The data approximate a Normal distribution. For

summer of 1816 (due to volcanic eruptions), which brought widespread crop failure throughout Europe, is apparent at the other tail of the distribution.

Assuming that variability will not change and that the normal approximation remains valid, we can estimate the change in the frequency of extreme events by shifting the distribution (as per Parry, 1985; Wigley, 1985; Mearns, et al., 1984). As shown in Figure 2.4b, the probability of the occurrence of very warm years increases dramatically as the mean temperature changes. For instance, in central England the probability of a summer like that of 1976 (or warmer) occurring in any given year increases from about 0.1% to 10%, a hundred-fold increase, by the year 2030. By 2050, such summers would occur every third year on average.

Of course the frequency of cool summers would decline. For instance, a cool summer like that of 1972 (or cooler) currently occurs about eight times a century, on average. By the year 2030, the probability of such events falls to 0.2%, that is, once in 500 years on average.

(a)

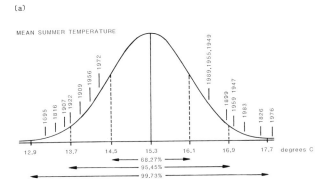

(b)

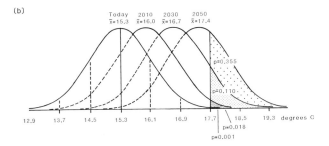

Figure 2.4 Central England summer mean temperatures: (a) Based on the period 1660–1989; (b) shifts in the probability distribution for 2010, 2030, 2050. Shaded areas denote changes in approximate exceedance probabilities for reference hot summer of 1976.

(a)

(b)

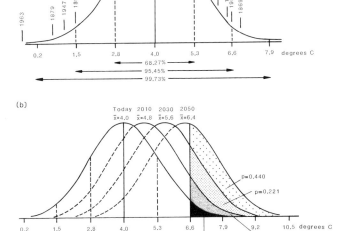

Figure 2.5 Central England winter mean temperatures: (a) Based on the period 1851–1989; (b) Shifts in the probability distribution for 2010, 2030 and 2050 according to average warming scenarios. Shaded areas indicate changes in exceedance probabilities for reference winter 1989.

## Winter

For the winter season (December, January, February), a southwest to northeast gradient of increasing temperature change is evident, which implies a trend toward more uniform average winter temperatures over the UK (Figure 2.3 a, c, & e). The mean winter temperature is projected to be 1.5–2.1°C warmer along this gradient by the year 2030. For southern England, for instance, this would imply a mean winter temperature comparable to that of Bordeaux, France.

Figure 2.5a shows the frequency distribution of mean winter temperature for central England with some examples of unusual winters, based on the period 1881–1989. As above, Figure 2.5b shows the shift in the distribution for the years 2010, 2030 and 2050 and the associated changes in probability for extreme events.

The winter of 1988/9 was indeed warm in central England. By 2030, the chances of a similar (or warmer) winter would be increased by a factor of ten, to 20% in any given year. In contrast, the very cold winters like 1962/3 and 1946/7 would have virtually no chance of occurrence.

## 2.5 THERMAL INDICES FOR IMPACT ASSESSMENT

The above scenarios of future temperature change can be further developed to provide thermal indices useful for impact assessment. One such simple index is based on the concept of 'degree-days': the number of degrees above (or below) a specific threshold temperature, accumulated over all days in the year in which the mean temperature is above (or below) the same threshold value. This thermal index can be employed in a variety of impact analyses.

To illustrate, maps of the percentage changes (in relation to the baseline averaging period, 1951–80 for a network of 162 temperature stations within the UK) in degree-days above 5°C and below 16°C are shown in figures 2.6 and 2.7 respectively. For degree-days above 5°C, which provides a rough measure of the thermal requirements for maturation of some crop varieties, the percentage changes implied by the time-dependent warming scenarios increase over time. As shown in Figure

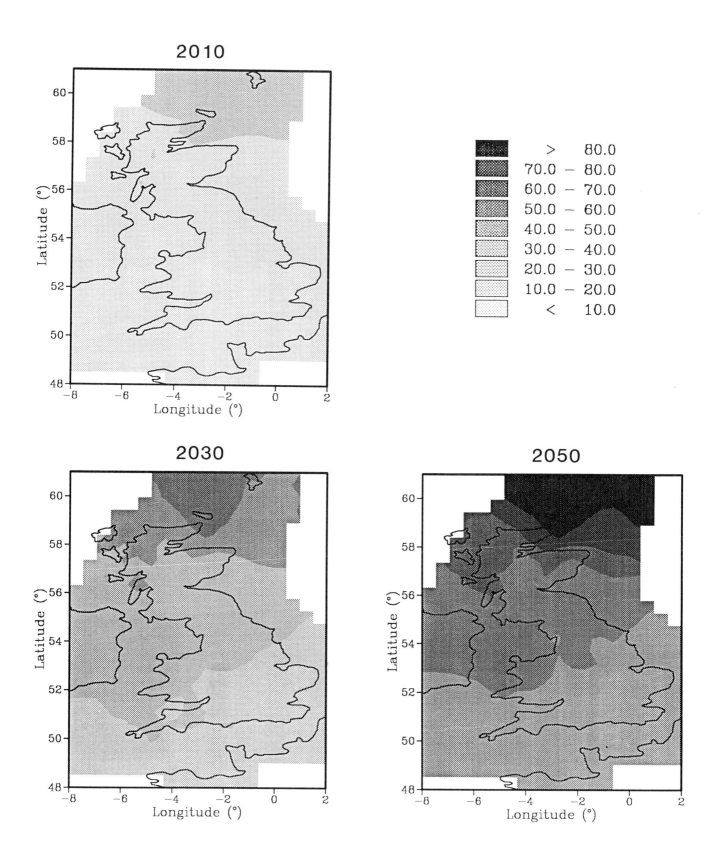

Figure 2.6    Percent increase in degree days above threshold temperature of 5°C.

2.6, the largest relative increases are in northern UK, because this is where the current baseline degree-days are least and the projected temperature changes are largest. By the year 2050 increases of 60–80% could be expected. This might imply, for example, a potential northward extension of the production area of some crops (e.g. maize) for which temperature is a limiting factor.

## 2010

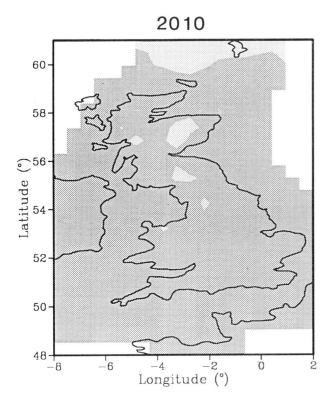

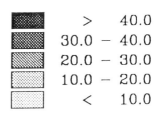

| | |
|---|---|
| | > 40.0 |
| | 30.0 – 40.0 |
| | 20.0 – 30.0 |
| | 10.0 – 20.0 |
| | < 10.0 |

## 2030

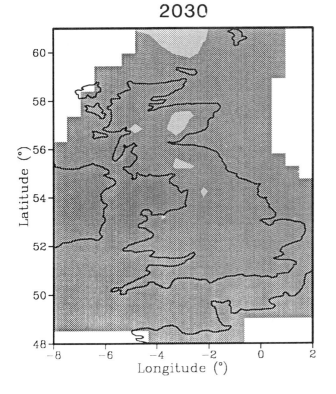

## 2050

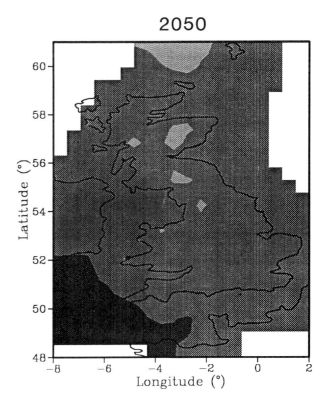

Figure 2.7    Percent decrease in degree days below threshold temperature of 16°C.

The percentage changes in accumulated degree-days below 16°C (an indicator of the energy requirements for space heating) is shown in Figure 2.7. In this case the number of degree-days decreases as the climate warms. The percentage decreases occur rather uniformly over the UK (the effect of larger warming in the north being offset by fewer baseline degree-days in the south). By the year 2050, decreases of 30–40% could be expected. This would imply that all Britons may require less energy for home heating.

## 2.6 PRECIPITATION CHANGES

From GCM results, in global-mean terms, a warmer world would be a wetter world. The percentage increases in global precipitation are directly related to the equilibrium temperature change for a $CO_2$ doubling, as shown by GCMs. The range of precipitation increase for a $CO_2$ doubling is from about 3–15%; roughly an 8% precipitation increase is associated with a global warming of 4°C (Mitchell, *et al.,* 1990).

The spatial patterns of precipitation changes predicted by GCMs are very unreliable. There are two major reasons for this conclusion. Firstly, most GCMs have serious difficulties in simulating the present precipitation patterns accurately at the regional scale. Secondly, the differences between GCMs for $CO_2$ experiments are often very large. This gives us little confidence in the results as predictions; they must be regarded useful only as scenarios. With these caveats in mind, the following model projections for the UK can be made.

*Summer*

The seasonal and annual equilibrium precipitation changes (in mm/day) for a single UK grid point as derived from five GCMs are shown in table 2.1. For the summer season two GCMs indicate an increase in precipitation and three indicate a decrease – a

**Table 2.1 Seasonal and annual precipitation anomalies (mm/day) for a grid point representation of UK from five GCM experiments for a doubling of $CO_2$ concentration. All points lie on the Greenwich meridian.**

| Model | Latitude | DJF | MAM | JJA | SON | ANN |
|---|---|---|---|---|---|---|
| UKMO | 52.5°N | 0.19 | 0.70 | 0.86 | 0.58 | 0.58 |
| OSU | 52.0°N | 0.18 | 0.01 | −0.06 | 0.06 | 0.05 |
| GFDL | 51.1°N | 0.50 | 0.26 | −0.39 | 0.00 | 0.09 |
| NCAR | 51.1°N | 0.82 | 1.37 | 0.22 | 0.75 | 0.51 |
| GISS | 50.87°N | 0.53 | 0.09 | −0.26 | 0.28 | 0.15 |
| Average | | 0.44 | 0.49 | 0.07 | 0.33 | 0.33 |

Source: Hulme and Jones, 1989

large range of uncertainty. The percentage change in precipitation (relative to the GCMs' simulations of present precipitation) from the average of individual model results is positive but close to zero (Wigley *et al.,* 1990). Geographically, the model-averaged results indicate that, although the uncertainties are large, the chances of drier summers in the future may be slightly higher in the south of the UK than in the north (Santer *et al.,* 1990).

These GCM results suggest that, for the equilibrium situation from a $CO_2$ doubling, it would be reasonable to assume 0% change in UK summer mean precipitation, with 'bounds' of ±30%. In order to derive time-dependent scenarios for the UK, this estimate can be scaled by the ratio $\triangle T/\triangle \overline{T}_{2x}$ (where $\triangle \overline{T}_{2x}$ is the average sensitivity of the GCMs to a $CO_2$ doubling, approximately 4°C). Using the best estimate of $\triangle T$ shown in Figure 2.1, the UK scenarios of summer precipitation change for 2010, 2030 and 2050 are 0±5%, 0±11% and 0±16% respectively.

It is important to bear in mind that a zero change in precipitation does not necessarily mean that UK summers will not become *drier*. With no change in precipitation, higher temperatures imply greater evaporation. All other factors being equal, this results in reduced soil moisture, runoff and streamflow, with possible secondary impacts on natural vegetation, agriculture, water resources, etc.

It is instructive to view the scenarios of precipitation change in the light of summer rainfall totals based on long term records. The distribution of England and Wales (E+W) area-averaged summer (JJA)

(a)

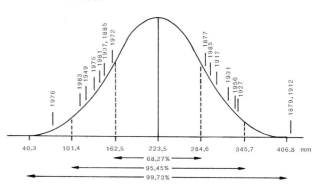

(b)

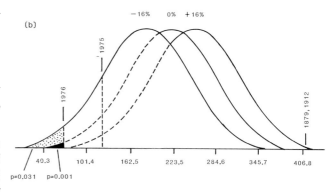

Figure 2.8 Summer precipitation for England and Wales: (a) Based on the period 1873–1987; (b) Shifts in the distribution for possible −16% and +16% change in average precipitation in 2050. Shaded area indicates changes in exceedance probabilities for reference drought of 1976.

precipitation (Wigley and Jones, 1987) for the period 1873–1987 is shown in Figure 2.8a. The data approximate a Normal distribution (with x=223.5 mm and S.D.=61.1 mm). Very wet or dry summers are noted for reference. For example, the summer of 1976, besides being the warmest summer on record (see above), was also the driest, with a low probability (p) of occurrence (p=0.007). At the 'wet tail', the very unusual rainy summers of 1879 and 1912 had nearly twice the expected rainfall (p=0.001).

By making the same assumptions as in Section 2.4 and shifting the distribution according to the scenarios of future change (as shown in Figure 2.8b), we can learn something about the possible sensitivity of extreme events to changes in mean precipitation. For example, if mean precipitation changes on average by as much as −16% by the year 2050, the chance of a dry (or drier) summer like 1976 occurring in any given year increases to p=0.031, i.e. about a fourfold increase (Table 2.2). On the other hand, a +16% change in precipitation would make the summer of 1976 exceedingly rare.

winter precipitation (relative to the GCMs' simulations of present UK precipitation) is about +15%, evenly distributed throughout the UK (Wigley et al., 1990; Santer et al., 1990). From Table 2.1, the range of uncertainty is roughly 0–30% increase.

Again, time-dependent scenarios of winter precipitation changes can be approximated as above. Thus for the years 2010, 2030 and 2050, the UK winter precipitation change scenarios are 3±3%, 5±5% and 8±8% respectively.

Such average precipitation changes imply a reduction in the chances of extreme dry winters. In recent decades, the dry winters of 1963/4 (p=0.014), 1962/3 (p=0.081) and 1975/6 (p=0.081) are notable. The occurrence of consecutive dry winters of 1962/3 and 1963/4 was very rare (p=0.001, assuming independent events). With the scenario of wetter winters the chances of such dry winters decline. For example, with the 'best estimate' scenario, the chances of a winter like that of 1963/4 would be reduced by half at the year 2050.

**Table 2.2 Exceedance probabilities of selected dry and wet summers for England and Wales. Baseline probabilities based on area-averaged precipitation data (Wigley and Jones, 1987) for the period 1873–1987. Estimates for years 2010, 2030 and 2050 assume simple shift in frequency distribution with no change in standard deviation, for the Business-As-Usual greenhouse-gas forcing scenario. The decimal places are for comparison only and do not represent precision.**

| Analogue Years | Baseline Probability (p) | Scenario: Year and ▲P (%) | | | | | |
|---|---|---|---|---|---|---|---|
| | | 2010 | | 2030 | | 2050 | |
| | | −5% | +5% | −11% | +11% | −16% | +16% |
| 1976 (dry) | 0.0071 | 0.0119 | 0.0043 | 0.0207 | 0.0022 | 0.0314 | 0.0012 |
| 1975–76 (dry)* | 0.0005 | 0.0013 | 0.0000 | 0.0032 | 0.0000 | 0.0063 | 0.0000 |
| 1912, 1879 (wet) | 0.0012 | 0.0006 | 0.0021 | 0.0006 | 0.0041 | 0.0001 | 0.0069 |

* probability of consecutive occurrence assuming independent events

One unusual feature of the 1976 summer was that it was preceded by the dry summer of 1975. Assuming summers to be independent of one another with respect to precipitation totals, the chance of the sequence of 1975–76 was very rare indeed, occurring on average once every two thousand years. With the 16% drier scenario, the 1975–76 sequence is twelve times more likely to occur in 2050; with the wetter scenario, the chances become negligible (Table 2.2).

*Winter*

For UK winters (and spring and autumn, as well as the annual total) GCMs agree that warmer is wetter. However, the equilibrium changes in precipitation rates predicted by GCMs for a $CO_2$ doubling vary considerably (Table 2.1). From the average of individual model results, the percentage change in

## 2.7 OTHER CLIMATE PARAMETERS

Unfortunately, as yet, very little can be said about future changes in such climate variables as precipitation intensity, cloudiness, windiness or storminess in the UK. For these, much depends on the regional details of changes in the general circulation and how they will affect the incidence of anticyclonic conditions (including blocking), storm tracks, etc., at the regional scale. At this scale, the uncertainty in climate modelling is very large indeed.

It should be borne in mind that there are some important parameters which help give the UK its characteristic climate and which will not change with global warming. For example, it is sometimes incorrectly inferred that since average temperatures in the UK may approach those of, say, Spain or Italy, the UK can look forward to a Mediterranean climate. This is not so. The winter sun in the UK will continue to follow the low geometrical paths it does at present, and the horizontal global radiation will remain relatively low in winter due to the high latitude of the UK. It is unlikely that the UK would be dominated by sub-tropical high pressure in the summer causing a winter-wet/summer-dry climate, as in the Mediterranean. Climate change is in store for the UK, but with an overall pattern that, in fact, may be unique, without a geographical analogue.

## 2.8 SEA LEVEL RISE

As the world warms, global-mean sea level is expected to rise due to thermal expansion of the oceans and to increased melting of mountain glaciers and the Greenland ice sheet. The most likely response of the Antarctic ice sheet is increased accumulation, contributing to a slight decrease in sea level.

Figure 2.9 shows the global sea level rise projections associated with the projected global warming. These projections include both the climate uncertainties, as reflected in the range of climate sensitivity values, as well as uncertainties associated with the factors contributing to future sea level rise. All projections are for the single 'Business-As-Usual' forcing scenario.

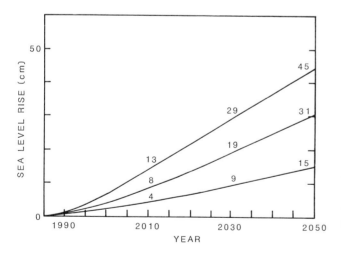

Figure 2.9  Projected global mean sea level rise (cm), 1985–2050, based on Business-As-Usual temperature projections. (Model results courtesy of S.C.B. Raper, Climatic Research Unit, UEA)

By the year 2030, global sea level is estimated to be approximately 20±10 cm higher than today. By 2050, the rise would be about 30±15 cm. These results are nearly identical to those found in the IPCC report (Warrick and Oerlemans et al., 1990), but more by coincidence than design (see Footnote 2). They are also consistent with other recent work (e.g. Raper et al., 1991; Warrick and Farmer, 1990; Oerlemans, 1989).

For the UK, the implications of a global sea level rise have to be considered in light of present regional trends in relative sea level (see Section 6). In general the southeast UK is sinking and the northwest UK is slowly rising, due to lingering crustal effects from the last glaciation. Thus in the south, a global sea level rise will exacerbate the problems of erosion, flooding and salination already being experienced. Local land movements also influence relative sea level. All of these factors have to be included in assessing the additional effects of a global sea level rise.

## 2.9 CLIMATE CHANGE COMMITMENT

Because of the lag in warming due to the thermal inertia of the oceans, additional future global warming is 'in the pipeline' *even if further increases in greenhouse-gas concentrations were to suddenly cease.* This additional warming, or 'warming commitment', can be defined as the eventual equilibrium warming minus the observed warming at any given time. For example, assuming a range of climate sensitivity of 2.0–4.0°C, an observed global warming over the last 100 years of 0.5°C (which in reality may not be wholly due to greenhouse forcing) and the current equivalent $CO_2$ concentration of 412ppmv, today's global warming commitment is 0.6–1.7°C. This should be the warming commitment for the UK summer as well, following our assumptions above.

Of course, since changes in the whole climate system are driven by temperature change induced by radiatively-active greenhouse gases, there are other 'commitments' as well – in precipitation and sea level, for example. The sea level commitment is particularly large. This is primarily because heat is redistributed within the oceans and thermal expansion continues to occur long after greenhouse-gas forcing stops, and because glaciers and ice sheets continue to respond to ongoing changes in temperature.

These 'commitments' mean that some future climate and sea level change is unavoidable – even with very stringent efforts to reduce greenhouse gas emissions. In the short-term (3–5 decades from now) very little greenhouse-gas-induced climate and sea level change projected under the Business-As-Usual scenario can be prevented. In short, it is likely that the UK will have to learn to live with climate change.

## 2.10 SUMMARY AND CONCLUSIONS

In this Section scenarios (*not* predictions) of UK climate change in the years 2010, 2030 and 2050 have been presented. This was accomplished by using time-dependent results from simple transient climate models to scale the spatial patterns of equilibrium warming as derived from averaging recent GCM results. A Business-As-Usual scenario of greenhouse-gas emissions was assumed. It was found that:

(i) At a global scale, the world will be 0.7–2.0°C warmer than today in the year 2030, with a best-estimate of 1.4°C.

(ii) For the summer season in the UK, the best-estimate is that, on average, the temperature change will be comparable to that of the global-mean value.

(iii) By 2030 the UK may experience mean winter temperatures about 1.5–2.1°C warmer than today, with the largest increases in northern areas.

(iv) The probability of occurrence of extreme warm years could increase dramatically. For instance the chances of a hot summer like 1976 could increase a hundredfold by 2030.

(v) Precipitation in the UK is most likely to increase in winter. A best-estimate is that, on average, winter precipitation will be 5% higher in 2030.

(vi) For summer precipitation changes, there is large uncertainty. Climate model results differ in direction as well as amount. A best-estimate is probably no change (with an uncertainty of ±11% by 2030), which would imply drier summers due to higher evaporation.

(vii) Global-mean sea level is estimated to be about 20 cm higher than today by the year 2030. For the UK, this estimate has to be adjusted for ongoing vertical land movements at local and regional scales.

(viii) Even if stringent measures are taken to reduce greenhouse-gas emissions, substantial changes in climate and sea level, such as those noted above, will occur anyway – a 'climate change commitment'. This is due to changes in greenhouse-gas concentrations that have already occurred and to lags in the climate system.

*It should be emphasized that the above estimates are scenarios, not predictions.* The uncertainties in model-based projections of regional climate changes are high. Future refinements in modelling, particularly in fully coupled ocean-atmosphere GCMs, are required in order to obtain realistic, time-dependent projections of climate change at regional scales.

### FOOTNOTES

1. The five models and the published results used were: Goddard Institute of Space Studies (GISS) (Hansen *et al.,* 1984); National Center for Atmospheric Research (NCAR) (Washington and Meehl, 1984); Geophysical Fluid Dynamics Laboratory (GFDL) (Wetherald and Manabe, 1986); United Kingdom Meteorological Office (UKMO) (Wilson and Mitchell, 1987); Oregon State University (OSU) (Schlesinger and Zhao, 1989). See Wigley *et al.,* (1990) for further description of method and results.

2. The IPCC adopted a B-A-U forcing scenario with an equivalent $CO_2$ doubling date of about 2020, a higher forcing than that used here. The IPCC range for $\triangle T_{2x}$ is 1.5–4.5°C, with a best estimate of 2.5°C. Finally, in calculating the 1990 future warming, IPCC takes the difference between the *model-predicted* $\triangle T$ by 1990 and the future $\triangle T$, whereas in this report the *observed* $\triangle T$ by 1990 is used (assumed to be 0.5°C). With the B-A-U scenario, the IPCC best estimate is that the world will be 0.5°C, 1.1°C and 1.7°C warmer than today by the years 2010, 2030 and 2050 respectively.

# Soils

## 3

**SUMMARY**

- The amount and availability of water stored in the soil is critical in determining land use. The water holding capacity of soils is likely to decrease and soil moisture deficits increase with higher temperatures. These changes would have a major effect on the types of crops, trees or other vegetation that soils in a particular area can support. The whole pattern of land use in the UK may change as a result.

- Changes in soil moisture content have implications for the workability of land. Cultivation may become more difficult under drier conditions and new types of machinery may be needed. The period in which the soil is easily cultivable may also change, with implications for sowing and other practices associated with crop production.

- Given drier, warmer summers, many soils would shrink more extensively than hitherto, and in winter under wetter conditions, would swell again. This alternate shrinkage and swelling would have important implications for the stability of foundations of buildings and other structures. The areas most affected would be central, eastern and southern England, where there are clay soils with a large shrink-swell potential.

- Soil shrinkage under drier warmer summers would lead to the formation of cracks linking the soil surface to underlying parent material. Such cracks would become the main pathways for movement of water, solutes and pollutants from the soil surface to depth with major effects on soil moisture balance, efficiency of fertilizers, herbicides and pesticides, and the rates at which water or pollutants reach watercourses.

- Some soils will change rapidly. The rate of loss of organic soils would increase significantly, affecting ecological habitats nationally and agriculture use in southern England. Currently poorly drained mineral soils may become drier and hence less of a problem whereas, close to low-lying coastal areas, well-drained soils may become poorly drained.

- Loss of organic matter and a decrease of soil biological diversity would reduce the stability of soil structure, resulting in increased vulnerability to land degradation.

- If water tables rose in response to a rise in sea level, a change in soil processes would result. Existing land drainage schemes may become ineffective and there would be a need for new drainage schemes if land is to remain suitable for arable cropping.

- Soils affected by rising sea levels would become saline and this would affect the types of crop that can be grown, as well as land management and the vulnerability of subsoil components of engineering structures to corrosion.

- Although outside the remit of this Review, it should be emphasised that soils represent a significant source and sink for greenhouse gases. Climate change is likely to alter the balance between the source and sink roles in such a way as to lead to a significant increase in the release of these gases.

## 3.1 INTRODUCTION AND BACKGROUND

Little attention has so far been given to the effects of climate change on soils, yet they, together with water, represent our most vital natural resource. Soils perform a number of essential functions including: the principal medium for plant growth, a foundation for buildings, a filter controlling water quality and flow rate, a source and sink for pollutants, and a substrate for fauna and flora. Soils are, therefore, of major importance for the stability of global processes and are one of the major buffers against man-induced climate change.

Climate is one of the main factors influencing soil formation and is responsible in part for the fact that soils change from region to region and country to country. It is the driving force behind most soil processes and is the dominant factor influencing the potential of soils for different uses. A change in climate, if maintained over at least decades, will lead to changes in soil properties and soil types. The extent and direction of these changes will vary with soil type.

Soils themselves are subject to changes but they can also influence climate change. Under the changes of climate described in Section 2, a number of soil properties will be liable to change, affecting the ability of the soil to grow particular crops or trees and altering productivity. Two of the most important such properties are organic matter content and waterholding capacity. An increase in temperature without concomitant increase in rainfall is likely to be reflected in increasing soil droughtiness and a decrease in soil stability. The role of soil as a filter controlling water quality and flow rate will change. Land degradation, particularly soil erosion, will increase as will soil shrinkage, the latter being of particular significance to the stability of buildings, reservoirs, etc.

Soils can influence climate change because they represent a source and sink for greenhouse gases. Soils are a major pool of carbon with an estimate of $2000 \times 10^{15}$ gC stored, mainly in topsoil (Bouwman, 1990). Significant quantities of nitrogen are also held. Climate change is likely to alter the balance between the source and sink roles of soils in such a way as to lead to a significant increase in release of greenhouse gases. Thus it is likely that a reduction of the soil carbon pool will accelerate the increase in atmospheric carbon dioxide which in turn leads to warming and to further changes in the soil pool. Methane is also an important component of wetland soils. Globally the emission of $CH_4$ from wetlands ranges from 40 to $160 \times 10^3$ g per year. It is suggested that by the year 2080 these current emissions will have doubled (Burke et al. 1990). This particular role of soils lies outside the remit of this Review Group.

Sea level change will have an important effect on soils. Obviously, low-lying areas will be more prone to flooding and current soil drainage systems could become ineffective. Water levels in low-lying areas may rise causing waterlogging and salination of soils, thus affecting the workability and trafficability of land, the types of crops that can be grown and the corrosion of civil engineering structures.

The most important climatic parameters affecting soils and their use are temperature and rainfall, (particularly amounts of rainfall, its distribution and intensity). The fact that reliable estimates cannot yet be made about rainfall is a major hindrance in predicting the effects of climate change on soils. In addition to temperature and rainfall, windspeed, frost days and radiation also affect soils.

The climate changes currently projected to occur in the UK will not be deleterious to all soils in all regions. Some soils, particularly those in the north and west, may improve and become more suited to current UK crops than hitherto. Too few in-depth investigations have been made to permit informed assessments of the main changes. The current body of knowledge is summarised in the following section.

## 3.2 ESTIMATED EFFECTS OF CLIMATE CHANGE AND SEA LEVEL RISE

### 3.2.1 Effects of climate change

Soils generally take several hundreds, if not thousands, of years to form. Under current temperature climatic conditions a crude estimate for the formation of new soil each year is 1 tonne/ha. The rate of soil formation can be expected to increase as the temperature increases, and particularly if this is combined with an increase in rainfall. Soil processes that are likely to be affected by an increase in soil temperature include chemical reactions in solution, diffusion-controlled reactions, and dissolution of solid and gaseous phases. Although some processes take centuries or more to make a distinct imprint, there are some important ones which operate over a 1-10 year time scale and will certainly be affected by the changes in climate projected between now and 2050. These will result in changes in organic matter content, waterholding capacity, shrinkage and structural stability. In addition, changes in climate could have a major impact on soil biolology (e.g. earthworms, microorganisms), affect the impact of soil-borne pests and diseases and influence the rate of flow of water and would-be pollutants through the soil.

## Organic matter

Organic matter, as well as being an important potential source of greenhouse gases, is an extremely important soil component. It provides the key to nutrient cycling, soil structure and water movement, soil stability, water-holding capacity, adsorption of pollutants, and acts as a substrate for microorganisms. It is critical in determining ecosystem responses to climate change. Of all the soil components, organic matter is the most vulnerable to increasing temperature and any decrease in rainfall. The rate of change in the organic matter content depends on the balance between the amount produced, which could increase in response to increased $CO_2$ in the atmosphere, increasing vegetation production, and decomposition, the rate of which will increase in response to increasing temperature. It is the differential response of these two processes which will dictate the long-term overall changes in soil organic matter but which is very difficult to predict until further research is done. Models have been developed relating to the effects of temperature, soil water content, carbon inputs and cultivation to organic matter content and turnover (e.g. Veroney et al. 1981; Molina et al., 1983; Sanchez and Miller, 1986; Parton et al., 1987; Jenkinson et al., 1987; Jenkinson, in press;) and these provide a useful basis for research into the effects of climate change.

Initial work carried out by Dr. P. Ineson (1990) in Meathop Wood, Cumbria, using calculations of current evapotranspiration and predicted evapotranspiration for 2050 based on a 3°C temperature increase and a 10% rainfall increase, indicates enhanced losses of organic matter. The scenario used is at the wetter end of the range predicted in Section 2 and the results point to significant losses of organic matter, particularly if the rainfall declines. The consequences of this reduction in organic matter include:

- A decrease in moisture holding capacity so that soils will become more droughty.
- A decrease in soil stability and increasing vulnerability to erosion and degradation.
- A decrease in the ability of soils to adsorb pollutants with implications for water quality.
- Increased mineralization leading to nitrate leaching with cost implications for the water supply industry.

- A possible decline in soil structure leading to increased run-off of water rather than its infiltration.
- A change in the soil's capacity to store and make available nutrients for crops, trees or other habitats.
- A loss of substrate for soil microorganisms and soil fauna generally, leading to a decline in diversity and population.

Over 5 per cent of the UK consists of peat soils with organic matter contents of 40-95%. With an increase in average annual temperature of even 1°C, as projected by 2030, this volume of peat will decrease with increased decomposition, releasing $CO_2$, $CH_4$ and $N_2O$ to the atmosphere. In addition to the peat soils, over 10% of the UK soils contain more than 7% organic matter and most of the others have topsoils with 2% or more organic matter. In all of these soils the organic matter, at least in the upper parts of the soil, has a dominant influence and will be the component most responsive to climate change.

## Water holding capacity

The interaction between climate and soil is the key to the suitability of an area of land for a particular crop, species of tree or ecological habitat. Especially critical is the amount and availability of water stored in the soil. Existing models developed for predicting the annual and seasonal soil moisture deficits and soil-available water for particular crops in the UK (Jones and Thomasson 1985, 1987) can be used to predict changes in these parameters with increasing temperature and/or rainfall. Moisture deficits are likely to increase with an increase in temperature in the absence of an increase in rainfall (Table 3.1) and may do so for small increases in rainfall. Spatial changes in maximum potential soil moisture deficit (PSMD), with the addition to the mean (Figure 3.1) of half the standard deviation (Figure 3.2) and one

**Table 3.1 Changes in soil moisture deficit, droughtiness and field capacity days with change in climate**

|  | Increased temperature/ decreased rainfall | Increased temperature/ no change in rainfall | Increased temperature/ increase* in rainfall |
|---|---|---|---|
| Moisture deficit | + | + | (−) |
| Droughtiness | + | + | (−) |
| Field capacity days | − | − | (+) |

* The extent of changes in this column will depend on the extent to which increased evapotranspiration offsets the increase in rainfall.

17

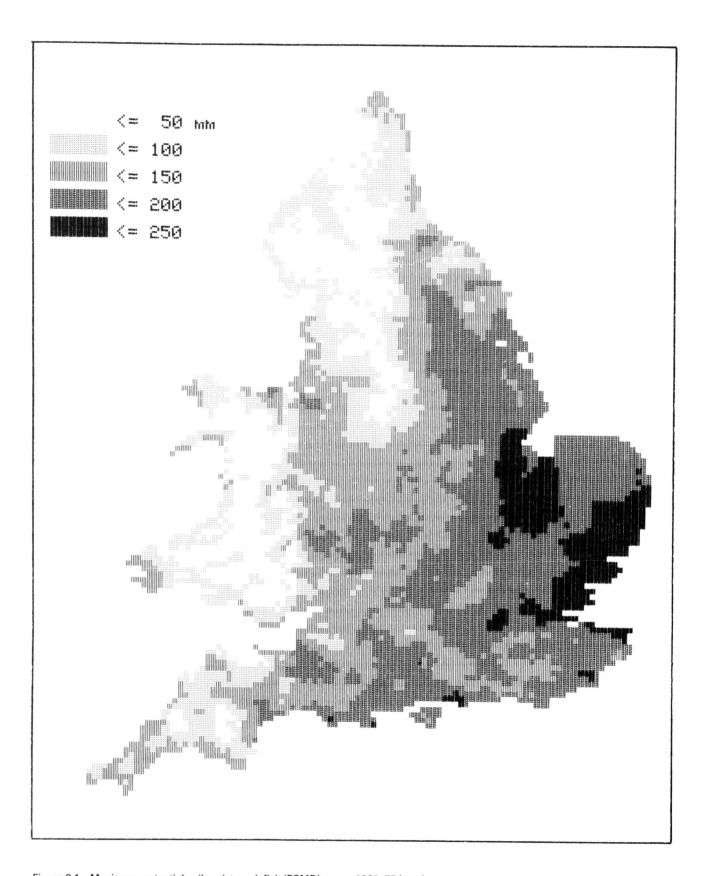

Figure 3.1   Maximum potential soil moisture deficit (PSMD), mean 1961–75 (mm).

standard deviation (Figure 3.3) are significant. The droughty year of 1975 was more extreme than the addition of one standard deviation to the mean, which for eastern England adds about 75 mm to the maximum PSMD. Yields of most agricultural crops were adversely affected as a result of drought in 1975 and 1976; for crops such as potatoes the effect was more severe than for cereals and was catastro-

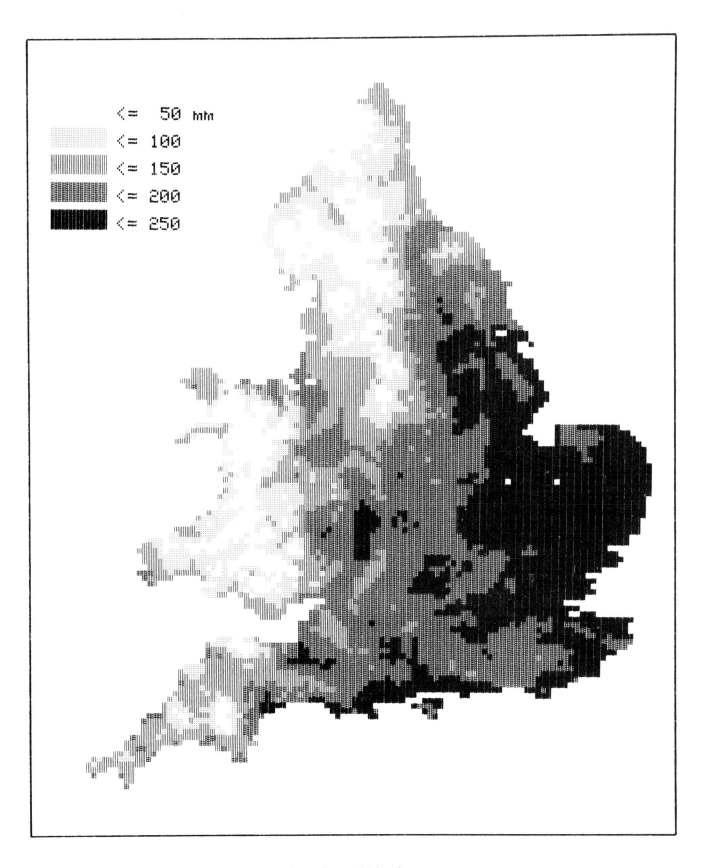

Figure 3.2  Maximum potential soil moisture deficit (PSMD), +0.5 SD (mm).

phic in the absence of irrigation. Subject to this degree of droughtiness on average, soils currently assessed as well suited to cereal production would become only moderately or marginally suited.

Sandy soils, which occupy about 5% of the UK, would become unsuitable for a wide range of crops without irrigation, particularly in the Midlands and eastern England.

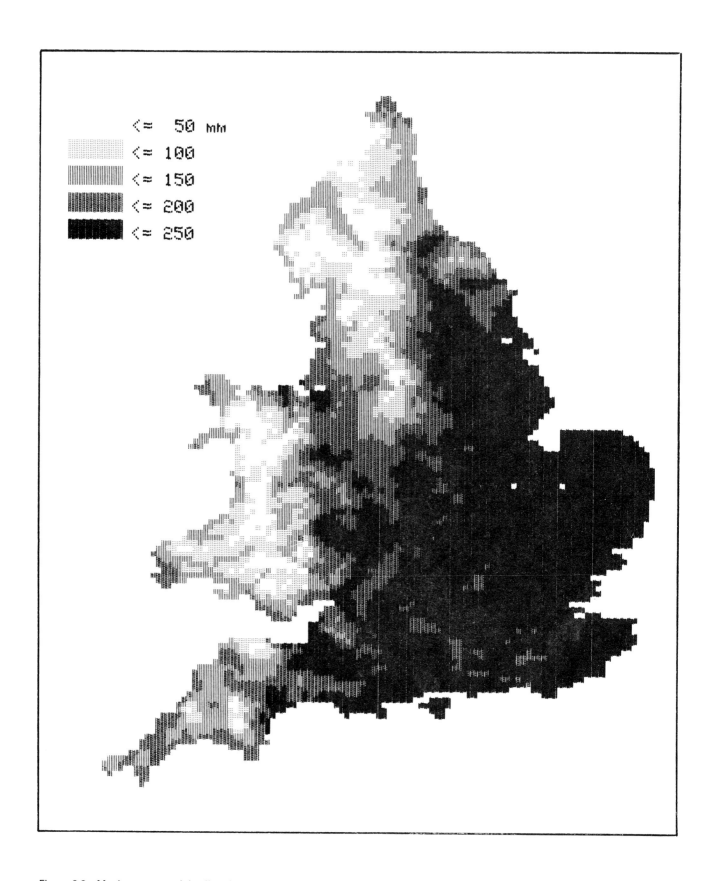

Figure 3.3  Maximum potential soil moisture deficit (PSMD), +1 SD (mm).

The period over which the soil is at field capacity in the UK is important in assessing the flexibility of land for cropping and the period over which crops can be sown, fertilized and harvested without damaging the soil (Thomasson and Jones, 1989). With a change in climate, the start and end of field capacity is likely to change considerably. Except on soils with a high shrink-swell potential, an increase in temperature is likely to decrease the number of days when the soil is at field capacity (Table 3.1) and hence provide the farmer with more opportunities for working on the land.

The significance of changes in the moisture status of soils in response to climate change is considerable, particularly for agriculture and forestry (see Section 5):

- The range of many species could move north-wards.
- The potential cultivable area would increase because some areas of the uplands would become cultivable.
- New crops, more suited to drier, warmer condi-tions could become widespread in southern England.
- The suitability of soils for particular species of crops or trees would change.

### Soil Shrinkage

Most clayey soils have the potential to swell when they are wet but shrink when dry. A soil's shrink-swell potential is a function of particular clay min-erals, mainly smectites. Many UK soils have high shrink-swell potential but under current climatic conditions rarely reach this potential. In the droughty year of 1976, such soils came close to achieving their shrinkage potential, and extensive subsidence damage was caused (Boden and Dri-scoll, 1987). Similar problems occurred in 1989 and 1990. A recent BRE Digest (BRE, 1990) shows that the cost of claims associated with subsidence and heave was £220m for 1989 and is predicted to be £400m in 1990. The vast majority of these claims will be associated with the droughty conditions. Given an increase in temperature similar to that predicted for climate change, there would be much more extreme drying out of the soil and maximum shrin-kage realised. The soils would crack extensively. The implications of this increase in shrinkage are considerable and have major economic conse-quences.

Many buildings erected on clay soils in central, eastern and southern England would be at risk from structural damage. During maximum shrinkage there would be subsidence and cracks likely to appear in buildings and roads. The distribution of clayey soils in England and Wales is shown in Figure 3.4; soils in Scotland are likely to be less affected because few contain significant amounts of smectite, although given extreme drying some shrinkage would take place.

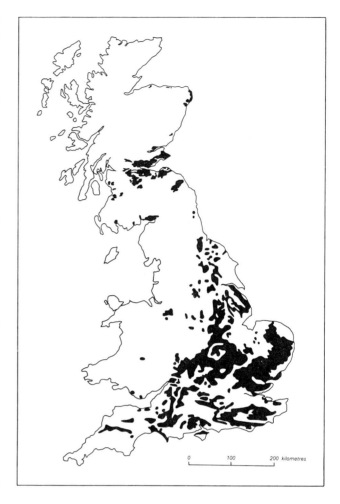

Figure 3.4   The distribution of clayey soils in Great Britain.

Associated with shrinkage is the development and enhancement of cracks linking the soil surface to the underlying parent material. In clayey soils in parti-cular, water falling on the surface passes preferen-tially through these cracks, termed 'by-pass' flow, with the following implications:

- Much of the water capable of wetting up the soil and providing a moisture reserve for crops, trees and other habitats by-passes the soil and is lost to the soil over the depth at which most plants root. This will have a major effect on the soil-water balance.

- There would be a much quicker response time to recharging of water courses. Instead of the soil acting as a filter, the rainwater would pass rapidly through cracks to join the water courses. The propensity to flash floods would be much increased (Section 6). Would-be pollutants applied at the soil surface would also have easier access to water courses.

- Fertilizers applied for crops and pesticides and herbicides directed to pests and weeds may have only limited effect because they may be removed by by-pass flow. By-pass drainage may also result in considerable losses of major nutrients, for example potassium and phosphate, as occurs from grazed pastures in drier areas of the world.

- The productivity of shrink-swell soils would decline as seasonal contrasts between shrinkage and swelling becomes more accentuated. Working the land when the soil is in the swollen condition is very difficult and cultivation when the soil is in its most shrunken and cracked state is also a problem.

Globally there are large areas of soils that shrink and swell strongly, known as Vertisols. They are recognised as one of the major problem soils of the world. As the average annual temperature increases, about 10% of UK soils would develop vertic properties.

Peat soils show some properties analogous to these shrinking clay soils in that they also shrink and on drying can become hydrophobic. Re-wetting of peat soils is slower than drying and may be irreversible. Particularly in the uplands of Scotland, Wales and northern England, this has consequences for increased run-off since the peat loses its spongy character and is then less able to act as a controler for water flow from the uplands.

*Land degradation*

A considerable area of the UK is potentially at risk

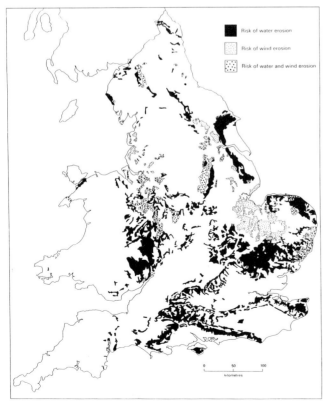

Figure 3.5   Land at most risk from erosion.

from soil erosion (Figure 3.5). The fact that this potential is currently only realised in some areas is due mainly to the fact that the present climate is not generally conducive to erosion. Land degradation may increase as a result of increasing temperature, consequent loss of organic matter and a possible decrease in biological diversity. This is suggested by comparison of land degradation in countries currently with a climate such as that projected for the UK in 2050. If the temperature changes are accompanied by decreased vegetation cover and a change in rainfall pattern to one with more periodic, higher intensity rainfall, then the amount of erosion could become catastrophic. The implications for agriculture, water quality and the construction industry would be important.

Although currently in the UK wind erosion is reasonably under control, light sandy soils and peat soils are particularly prone to drying out. So far the models used for predicting climate change can provide little information on changes in wind intensity, but in the event of winds increasing in speed and/or occurring when these soils are dry and without vegetation, erosion of these soils would be a major problem.

### 3.2.2.  Effects of sea level rise

Sea level rise will lead to the flooding of areas of good quality soils currently supporting agriculture or important ecological habitats (Boorman *et al.*, 1989). Slightly higher land beyond the flooded areas will also be affected. Areas such as the Fens (long considered to have some of the most fertile soils of the UK), around the Wash, along coastal valleys of East Anglia, the Halvergate Marshes and North Kent Marshes will all be affected.

Of the 8.15 million ha. of Grade 1–3 agricultural land in England and Wales 8% lies below the 5 metre contour and hence is vulnerable with respect to sea level rise (Whittle, 1990). In particular this 8% includes 198,000 ha. (57%) of the Grade I land in England and Wales, land which has the highest production potential and usually the most flexibility.

Sea level rise, with higher water table and increasing frequency of flood events, is likely to lead to downgrading of this high quality land by at least two grades, to 3 or 4. Some of the potential cost implications of this are outlined in Table 3.2.

There are three principal implications for UK soils of a rise in sea level similar to that currently projected to occur and described in Section 2: loss of marsh soils in some areas and increase in their occurrence in other areas, shallow water-table and, hence, poor drainage and salinity.

**Table 3.2 Some generalised economic implications of sea level rise for agricultural use of land**

| | Grade 1 (£/ha) | Grade 3 (£/ha) | Amounts involved in changes of Grade 1 Land below 5m (£ million) |
|---|---|---|---|
| Gross Margins-Winter Wheat* | 605 | 475 | 25.7 |
| Gross Margins-Sugar Beet* | 1075 | 795 | 55.4 |
| Land Prices-Grade 1 Fenland | 7500 | 4000 | 69.3 |
| Land Rental | 159 | 103 | 11.1 |

*Information taken from Nix (1990) with assumption that Grade 3 land corresponds to 'Average Productivity'.

*Marsh soils*

Some marsh soils will be permanently flooded. Other land which is flooded to shallow depth will develop into marshland with poorly drained soils subject to anaerobic conditions. If free of sulphate the new areas of flooded land developing into marsh soils would release methane through microbial decomposition of the organic matter and nitrous oxide by reduction of any nitrate formed. Organic matter accumulating from decaying marsh vegetation would be in contact with brackish water conditions, providing ideal conditions for the formation of potential acid sulphate soils. Soil organic matter in combination with sulphate from the brackish water results in the formation of pyrite and organic sulphur compounds. In the event of such soils being drained later, extremely acid conditions would result, preventing the use of the land for normal cropping.

*High water tables*

Rising sea level will cause river beds to aggrade and levees to increase in height along lower river courses, thus impeding drainage from interior basins. Water tables will rise and existing land drainage systems will become ineffective. New drainage systems will be required if the land is to be used for arable agriculture and land currently in agricultural use will be more expensive to maintain.

*Salinity*

Soils affected by rising sea level will become increasingly saline, not only through direct inundation but also through penetration of saline groundwater. Depending on the levels of salinity, and there is little information by which to determine quantitatively expected concentrations, the soils may only be suitable for salt-tolerant crops and vegetation. Salinity may be more of a problem on the better soils such as Fen peats or silts; because of their permeability, the salts will be readily distributed throughout the soil. Saline soils are difficult to manage effectively in agriculture because they are prone to structural problems (Hazelden *et al.*, 1986). Concrete, pipelines, cables etc., require much stronger protection in saline than in non-saline soils. Any new structures will need added protection against saline conditions.

## 3.3 UNCERTAINTIES AND UNKNOWNS

The uncertainty with respect to changes in rainfall amount and distribution between now and 2050 is a major drawback in assessing the impact of climate change on soils.

Because little research has so far been directed to investigating the impact of various scenarios of climate change on soils, many of the changes relating to soils can only be described in terms of an increase in temperature and an increase, status quo, or decrease in rainfall. Rates of change for soil properties can be predicted from a broad research base in soil properties.

The direction in which a particular soil change is likely to go can be identified with some assurance, the main exception being where an increase in temperature combines with an increase in rainfall. In this case, it is difficult to ascertain the likelihood of increased evapotranspiration negating the effects of increased rainwater penetrating the soil. It is difficult to estimate the rate of increase in land degradation because it is dependent on a range of soil properties (which themselves are undergoing change) and on land use response, especially vegetation cover at critical periods. The soundest way of proceeding may be to relate to analogous situations in those countries already subject to the climate which is projected for the UK and compare these assessments with analyses of changes expected in British soils based on fundamental relationships. This, however, provides little information relevant to rates of change affecting soils and this is an important gap.

Some soils will improve, others will deteriorate, particularly with respect to current land use. Insufficient research has so far been done to identify on a UK basis those soils that are likely to become more suited to certain crops and other uses, and those that will become less suited.

## 3.4 PRINCIPAL IMPLICATIONS

Given drier, warmer summers, many soils will shrink more extensively than hitherto, with important implications for the stability of foundations of buildings and other structures. The areas most affected will be central, eastern and southern England where there are clayey soils with a large shrink-swell potential.

The waterholding capacity of soils is likely to decrease in response to a decrease in the content of organic matter. Soil moisture deficits will increase and these changes will have a major effect on the types of crops, trees or other land uses that the soils in a particular area can support. It cannot be assumed that shortfalls can be corrected by irrigation because there may be insufficient water to meet demand (Section 7). In addition the increased use of groundwater for irrigation would lower the water level and increase the likelihood of saline intrusion into freshwater supplies. The whole pattern of land use in the UK may change as a result of a decrease in water holding capacity.

Loss of organic matter and a possible decrease of biological diversity will reduce the stability of soil structure, resulting in increased vulnerability to land degradation.

Rising water tables in response to a rise in sea level will lead to a change in soil processes. Existing land drainage schemes may become ineffective and new drainage schemes will be needed to maintain the suitability of low-lying land for arable cropping.

Soils affected by rising sea levels will become saline and this will affect the types of crops that can be grown, land management and the vulnerability of engineering structures to corrosion.

## 3.5 RESEARCH AND POLICY NEEDS

### 3.5.1. Future research effort

Little research has so far been undertaken with respect to: soils as a net sink or source of greenhouse gases, the effects of climate change on soils and the interactions and feedback effects between vegetation and soil. So far, the key role of soils has not been recognised by those assessing the implications of climate change. In this respect soil science is behind a number of other disciplines.

Seven broad research needs can be identified

(i)  Changes in climate involving temperature, evaporation and rainfall will lead to changes in soil organic matter dynamics which need to be modelled, and the implications of such changes determined for the release of $CO_2$ and other greenhouse gases to the atmosphere and for soil properties that affect land use and its sustainability. Most discussion in this area has centred on changes in amounts and more attention needs to be given to fluxes of carbon and associated elements through the soil/plant/atmosphere system.

(ii)  Given a change in climate, soil processes will change, gradually leading to new soil types in the UK. Research will be needed to determine how the processes will change, what will be the rate of change and the implications of such changes for land use, land management and water quality. The change in the shrinkage capacity of soils needs to be given some emphasis. The research should draw on analogues which currently occur abroad under climates similar to those currently projected to occur in the UK in the future.

(iii)  There needs to be a much better understanding of the sensitivity of UK soils to climate change both with respect to the amount and rate of change of properties and also in relation to whether soils will decline or improve in quality for a particular land use.

(iv)  In the absence of subsidies and other forms of economic support intervention, the interaction between climate and soil governs land use. Models will need to be developed to predict the effects of climate change on existing and future land use. The rate of change of land use and soils in relation to the rate of climate change will need to be considered.

(v)  Climate change may well have a major impact on the sensitivity of land to degradation processes, such as erosion. Research will be needed to determine the types of land degradation that may be associated with various scenarios of climate change and the mechanisms by which they will be brought about.

(vi)  The effect of sea level rise on land drainage should be studied. There may be a significant increase in the area of soils affected by salinity, and its effect on soil-crop interaction and on stability of civil engineering structures needs to be researched.

(vii)  Technology needs to be developed to offset the deleterious effects of climate change on soils. Research in agricultural engineering should be promoted to examine ways of counteracting or managing shrinkage effects, land degradation, water management and salinity control.

### 3.5.2 Policy recommendations

In spite of the uncertainties described above, a number of policy initiatives are required. These include:

(i)   The development of a clear indication of the contribution of soils to enhancement of climate change through the significant release of greenhouse gases and to include this in global warming assessments.

(ii)  The review of land-use policy for the period to 2050 with respect to the changing potential of soils for crops, trees and other habitats, and to draw up guidelines for decisions on whether to protect certain soils and land or whether to sacrifice them.

(iii) The examination of current building structures with respect to their ability to withstand increased soil shrinkage and the development of recommendations for future civil engineering structures.

(iv)  The development of a soil conservation policy under which land degradation would be monitored, effects on soils, industry and water supply assessed and remedial measures established.

(v)   In view of the strong influence of soils on the flow regime of rivers and the quality of both these and ground waters, future water resource policy discussions need to take soils into account.

(vi)  Consideration needs to be given to the effect of increased soil salinity on the agricultural potential of land and on current civil engineering structures such as concrete buildings and pipeline.

# Flora, Fauna and Landscape 4

**SUMMARY**

- The natural biota are sensitive to changes in almost all aspects of climate and weather. The parameters to which they would be most sensitive are: extreme weather events or seasons, soil water deficits, changes in mean temperature and increases in atmospheric $CO_2$ concentration. Any sustained rise in mean temperature exceeding 1°C will have very significant effects on the UK flora and fauna.

- Where rainfall is plentiful, an increase in $CO_2$ concentration accompanied by a lengthening of the growing season is likely to increase the productivity of vegetation. High $CO_2$ levels would also alter plant development and water use in ways that are not yet understood.

- There may be significant movement of species northwards and to higher elevations. The rate of northward movement for the climatic scenarios given is potentially 100 km/decade although few species are capable of migrating at this rate. Insects (invertebrates), birds and 'weedy' plant species will move first. Species which do not extend their ranges will not necessarily be affected adversely – many species will adapt to climate change. However, the predicted rate of climate change is too rapid for many species (e.g. most trees) to adapt genetically.

- There will be an increased probability of invasion and spread of alien weeds, pests, diseases and viruses, some of which could be potentially harmful.

- Many native species and communities will be adversely affected and may be lost to the UK. The losses (of both plants and animals) will occur particularly within (i) montane communities, (ii) salt marsh and coastal communities, (iii) confined 'island' habitats, and (iv) wetlands and peatlands.

- An increase in sea level of 20–30 cm would have an impact on about 10% of notified nature reserves, and might adversely affect invertebrates and birds which inhabit mudflats.

- There may be an increase in the overall number of species of invertebrates, birds and mammals in much of the UK (following migration and invasion), but the natural flora is likely to be impoverished by the loss of many endangered species which occur in isolated, damp, coastal or cool habitats.

- Land-use changes brought about by changes in agricultural and forestry policy in the UK could influence the natural biota as much as climate change itself.

## 4.1 INTRODUCTION AND BACKGROUND

This section considers the likely effects on the UK flora and fauna (the biota) of the changes in atmospheric $CO_2$ concentrations and temperature estimated as likely to occur in the UK. It also considers the sensitivity of the UK biota to possible changes in rainfall and a greater frequency of extreme events such as droughts and storms, although there are currently no scenarios for change in these climatic variables. It does not consider the effects of the biota on the climate, and thus does not examine issues concerning carbon storage, the terrestrial carbon cycle and land-atmosphere exchange of methane and nitrous oxide.

The main sources of information for this section were a number of expert reviews and books that have recently summarised the best available information on the relationships between the UK biota and climate (Ford, 1978; Woodward, 1987; George, 1988; Grime and Callaghan, 1988; Cannell, et al., 1989; Eamus and Jarvis, 1989; Jarvis, 1989; Cannell and Hooper, 1990). Draft papers prepared by the IPCC were also consulted, as were reviews of impacts in other countries.

There are three features about the biota of the UK that must be borne in mind when considering the impacts of climate change.

First, the natural landscape of the UK has been profoundly altered by man. Most of the land surface of the UK is used for arable farming, grazing, forestry, mining, roads, housing or industrial development. Thus, the natural flora and fauna exists within a largely man-made landscape, and its response to climate change depends on the pattern of current land use and on the changes in land use that will occur in the future.

Second, the UK has a maritime climate with a small seasonal amplitude in temperature over the range that is critical for life. A small increase in mean temperature greatly extends the period when the temperature is above the threshold for plant growth. Also, the UK spans eleven degrees of latitude (50° to 61°N), over which there is a marked north-south gradient in summer temperatures and of sub-zero temperatures in winter. Because of this gradient, a large proportion of the flora and fauna have part of their northern limits in the UK, often coinciding with isotherms. Thus, much of the flora can be classified as being of either 'southern' or 'northern' type (Matthews, 1955). For example, 45 of the 57 UK butterfly species have northern limits in the UK, as do 19 of the 52 UK species of wild mammals. Consequently, there is considerable scope within the UK for expansion of species northwards following climatic warming.

Third, Britain is an island, and it has a relatively impoverished native flora and fauna, consisting largely of those species that invaded between about 10000 BP, when the land became free of ice, and 7000 BP, when the land connection with the continent of Europe was severed. The island status of Britain means that extinctions are more likely than on a large land mass, and, if our diversity of flora and fauna is to be preserved, special consideration has to be given to dispersal within the confines of the UK, and to invasions or introductions from overseas.

Ecologists in the UK are well placed to be able to develop models that might predict the effects of climate change on the native biota. The flora has been intensively mapped and studied (Perring and Walters, 1962; Grime, et al., 1988; Rodwell, 1990). There are also extensive historic phenological records of flora and fauna (Margary, 1926; Jeffree, 1960), and several studies on the effects of enhanced $CO_2$ levels (Jarvis, 1989). The Biological Records Centre of the Institute of Terrestrial Ecology (ITE) has data on 8500 species of plants and animals. The Rothamsted light trap (moths) and suction trap (aphids) records contain long-term information on invertebrate numbers, as does the Nature Conservancy Council (NCC) invertebrate site register, and the ITE/NCC butterfly record. Also, good distribution data for breeding and wintering birds have been collected by the British Trust for Ornithology, which has monitored the numbers of common species for over 25 years.

Information and hypotheses about the impacts of climate change on the biota of the UK have been obtained mainly by (i) analysing the current distributions of plants and animals in relation to climatic patterns, (ii) analysing historic changes in the flora and fauna in relation to changes in climate or short-term weather, and (iii) reasoning from ecophysiological understanding developed by experimentation in controlled environments and by observation in the field. More formal approaches, using conceptual models of plant and animal community responses need further development (see below).

## 4.2 ESTIMATED EFFECTS OF CLIMATE CHANGE AND SEA LEVEL RISE

Before considering the effects of climate change, it is important to address the issues of land-use change, and the sensitivity of the biota to climatic and weather variables.

## Land use

Future changes in the landscape of the UK, and changes in the species composition and distribution of plants and animals within that landscape, could be influenced as much, or more, by socioeconomic decisions on land use as by changes in the climate. Hodgson (1986a,b) showed that the major changes in land use that have occurred in the UK since 1930 have brought about rapid and continuing changes in the flora. For instance, in lowland Britain, there has been a major expansion in the range and abundance of fast-growing perennials with high fecundity, and of prolific ephemerals, largely as a result of intensive agriculture, and some of the upland moorland vegetation has been replaced by trees with distinctive changes in the ground vegetation and fauna. Any future changes in the use of arable land (for set-aside or woodlands, for instance), in the extent or nature of pastures and grazing, and in the extent and type of forestry, will have a major impact on the future flora and fauna. Some of the decisions on future land use in the UK may, themselves, be related to climate change, both here and in response to changes elsewhere in the world.

## Relative sensitivity to different climatic variables

Major impacts on the flora and fauna may be brought about by changes in the frequency or magnitude of extreme events, such as droughts (as in 1976), high winds (as in 1987 and 1990), fires, floods and unseasonal frosts. These events have had great impacts in the past, and are likely to be major agents of large scale mortalities, rapid shifts in species distributions, invasions and extinctions in the future.

The next most important change (if it occurred) would be an increase in the magnitude and duration of soil water deficits resulting from decreased rainfall and/or increased evaporation. Evapotranspiration is likely to be greater following an increase in air temperature, an extension of the growing season, and greater biomass production. Increased soil water deficits would have a large impact on the flora and fauna throughout the UK, in northern wetlands as well as southern England.

Changes in mean temperature are likely to become important once the increase in temperature exceeds about 0.5°C and has persisted for a decade or so. The main effects may result from (i) the absence of snow and reductions in periods of critical freezing or chilling temperatures in winter, (ii) an increase in the length of the growing season (which is about 14 days in England for a 0.5°C increase), and (iii) an increase in maximum summer temperatures which

will favour the reproduction and dispersal of many warmth-loving species. Some researchers consider that the present distributions of some plants and animals are determined more by mean climatic conditions than by the frequency of extreme weather events.

The responses of individual organisms to climatic warming depends critically on the magnitude of a set of temperature thresholds for their growth and development. These thresholds include (i) base and optimum temperatures for plant growth (0–6°C), (ii) threshold temperature conditions for emergence in the spring (of buds, insects, hibernating animals etc.), (iii) chilling requirements to release dormancy (duration of temperatures −2 to 8°C), (iv) threshold temperatures for flowering, pollen tube growth and seed production, and (v) developmental thresholds for invertebrates (e.g. for the production of several generations per year, and for winter survival). In most cases, critical threshold temperatures are poorly known, even for common species.

At present, it is not possible to evaluate the direct impacts on vegetation of the projected increase in atmospheric $CO_2$ levels to about 530 ppmv by 2050 AD, relative to the impacts of the projected accompanying increase in temperature of about 2°C. Effects of the 80 ppmv increase in $CO_2$ concentration since pre-industrial times have been masked by changes in land use, farming methods, the introduction of new cultivars, pollution, etc. However, a further increase in $CO_2$ levels of 200 ppmv could induce large differences among species in growth, water use and development that could be as important as the responses to temperature. There could also be important interactions between responses to increased $CO_2$ levels and temperature.

### 4.2.1 Effects of climate change

The main issues are the composition of existing communities, the migration and invasion of species, losses and extinctions, outbreaks of pests and diseases, productivity, and landscape changes.

### Changes in the composition of existing communities

Given the warming scenarios presented, noticeable changes may be expected in the composition of plant and animal communities.

The animal populations (their sizes and distributions) that may be most affected by climate changes are those at the margins of their distribution, where the populations are influenced more by climate than by population density. However, it is impossible to generalize; individual species among the invertebrates, cold blooded animals, birds, mammals and fish are likely to respond differently.

Amphibia and reptiles, (which are cold-blooded) will not necessarily benefit from warmer conditions. Their populations are already limited more by the availability of habitats than by climate; mild winters increase maintenance costs, disrupt hibernation when food is unavailable and favour disease; warm summers will not promote greater activity of species spread unless they are sunnier; and dry summers might dry out amphibian breeding sites.

Bird populations in the UK are in constant flux. The populations of many species often crash and rapidly recover (e.g. some passerines) or they fluctuate from year-to-year for complex reasons concerning their habitats and food supplies as well as the weather. Rainfall at critical times can be as important as temperature. However, climatic warming is likely to benefit southern species and those currently adversely affected by severe winters.

Changes in plant communities may be seen first in montane areas, and among species occupying damp, cool refugia, especially mosses. If southern Britain became drier, grasses and species that overwinter below ground might be favoured. Again, it is difficult to generalize; the changes within plant communities depend upon many factors influencing inter-species competition, reproductive success and the establishment of seedlings.

The growth and flowering of most plant species, and the activity and reproduction of many animal species, will occur earlier in the spring. Some species will benefit from this shift in phenology more than others. Plants with low DNA amounts (which currently grow in late spring and summer) may benefit at the expense of species with high DNA amounts (like bluebells and some forage grasses) which already grow in early spring. Species which require winter chilling to release them from dormancy, and whose spring activity is triggered by increasing daylengths, might be unable to take advantage of early springs and so become at a disadvantage.

Flowering and successful seed production is favoured in many plant species by high summer temperatures and some degree of water stress. Tree species may have more 'mast' years (as in 1990) and plants at the northern limit of their range may flower and produce more seed than at present.

High temperatures may increase the incidence of algal blooms in freshwaters which have high nutrient levels (as in 1990). Trout in southern waters, and arctic charr and whitefish in northern waters, might decline.

The projected increase in atmospheric $CO_2$ concentrations may also bring about changes in the competitive balance among plant species whose assimilation and water use respond differently to enhanced $CO_2$ (see below). The productivity of the C4 *Spartina* species will be increased by higher temperatures.

Many of the changes in communities may be subtle and gradual, altering the competitive balances among species, and producing some communities which are unstable, in which some species may die following extreme weather, especially if climate change is faster than rates of species migration.

*Migration\* and invasion of species.*

There is great potential for a northward movement of the many plant and animal species which have northern boundaries in the UK, and also for expansion of species to higher elevations. A number of continental species with a toehold in southwestern England may well expand their distribution northwards (Grime and Callaghan, 1988).

A shift of 1°C mean annual temperature in the UK is equivalent to a latitudinal shift of 200–300 km or an altitudinal shift of 150–200 m (Parry, *et al.,* 1989). The 'Business-As-Usual scenario' for warming in the UK (Section 2) is equivalent to a northward shift in growing day degrees of about 100 km per decade.

A significant element of the rare and protected flora and fauna of the UK has a northern and montane distribution; species within these communities are likely to retreat, and their survival in the UK may become threatened if they have nowhere to which to retreat.

Table 4.1 lists specific examples of the UK species or types of animals and plants that are likely to increase or decrease in abundance and/or range as a result of climatic warming.

Plant and animal communities will not move as a whole; some species will migrate faster than others, giving new assemblages.

The first organisms to migrate will be those that are most mobile, particularly the warmth-loving

---

\* The term migration is not used here in the strict sense employed by zoologists when referring to migrant species, but rather it is used to mean the spread of species and their colonization of new areas.

**Table 4.1 Examples of the UK flora and fauna that are likely to be 'winners' and 'losers' following climatic warming. 'Winning' means an increase in abundance and/or range, and 'losing' means a decrease in abundance and/or range.**

|  | 'Winners' | 'Losers' |
|---|---|---|
| Plants | Improved seed production in e.g. stemless thistle and small-leaved lime. Southern species of disturbed ground e.g. black twitch, wall barley and prickly lettuce. | Woodland geophytes such as bluebell and ramsons. Northern species such as Jacobs ladder and mossy saxifrage. |
| Invertebrates | Dragonflies. Aphids. Most butterflies and moths. | Midges. Alpine sawflies. |
| Birds | Dartford warbler, Stonechat. | Snow bunting, Dotterel. |
| Mammals | Many bat species. | Mountain hare. |
| Fish | Salmon, Carp. | Arctic char. Freshwater whitefish. |

invertebrates of productive, disturbed habitats (Southwood, 1962), but also some birds and mammals, where food supplies and habitats permit.

Invertebrates will be especially sensitive indicators of climate change. Species that overwinter as adults or larvae will benefit from the absence of temperatures below about −5°C and warm weather in spring will favour the survival, mating success and reproductive activity of many insect species. Many invertebrates, birds and some mammals which are currently confined to the south of England could spread northwards within a few years, and some species of invertebrates and birds that are mainly summer visitors may become resident and more abundant. Figure 4.1 shows one example: most

dragonfly species (Odonata) occur in the south of England where mean April temperatures exceed 9°C; a 1°C rise in temperature would enable many species to spread to northern England and southern Scotland.

The plants that will migrate first are likely to be the 'weedy' ephemeral species, with high fecundity and rapid dispersal, often species of agricultural land. The plants most resistant to movement will be those with a long lifespan and clonal plants, including trees, shrubs and many plants of unproductive heathland and grassland.

The migration of plants is likely to occur in spurts, following extreme events, like droughts, fires or

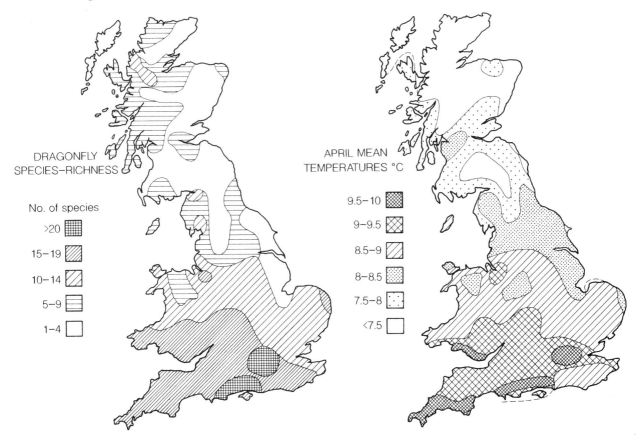

Figure 4.1  Number of dragonfly species (Odonata) found in Great Britain, and distribution of April mean temperature.

storms, which kill areas of vegetation and permit new species to colonize. Most species are unable to establish themselves from seed within existing closed communities. Similarly, the retreat of northerly species may occur following extreme events such as warm, dry summers.

As part of the process of plant and animal migration it is expected that some species (not whole communities) will invade from France and Spain, and many exotics from elsewhere could become established. These organisms are likely to include some Mediterranean plant species (some of which already have a 'toehold' in southern England), garden plants and captive animal species that may be able to survive and better compete with native species following climatic warming. The alien organisms are likely to include exotic weeds, pests, diseases and viruses, some of which could be harmful (see below). Alien plant species are already spreading to unexploited niches in productive farmland; this spread may be enhanced by climatic warming.

### Losses and extinctions

Rapid climate change is likely to accelerate the rate at which species are lost from particular regions or become extinct in the UK. Extinctions will often occur following extreme events.

The arrival and spread of 'new' or rare species of invertebrates and birds may well exceed the loss of species, so that in some parts of the UK the fauna may be enriched. However, the flora is more likely to become impoverished, because species-rich native communities could be replaced by a few aggressive species (like rhododendron) and many losses or extinctions could occur. A substantial number of 300 or so endangered plant species in the UK may be lost, although some rarities in the south may be favoured.

The main species or communities that may become endangered by climate change include the following:
- Montane/alpine and northerly/arctic plant and animal species which have 'nowhere to go', including whitefish, mountain hare, some bird species and many rare plant species.
- Species of salt marshes and 'soft' coastal communities that cannot retreat landward in response to sea level rise.
- Species confined to particular locations from which they cannot readily escape, such as cool, damp refugia on north-facing slopes, and species adapted to growing in infertile soils that are surrounded by fertile agricultural land, that is, species that cannot migrate because their habitats are fragmented and isolated.

- In the long term, long-lived, slowly reproducing or clonal plant species, including trees, which migrate very slowly and may decline as a result of increases in temperature (see Section 5).

### Pest and disease outbreaks

The climatic and biotic variables leading to pest and disease outbreaks are often very complex and difficult to predict, especially when they involve coupled predator-prey or herbivore-plant systems. Nevertheless, pest and disease outbreaks may become more likely as the climate warms, especially if the climate becomes more variable.

A rise in winter temperatures will give rise to an increase in the abundance of many aphid species, some of which are vectors of plant and animal viruses.

Plants that become stressed by drought or high temperatures are likely to become more susceptible to attack by certain insects, despite having lower internal nitrogen concentrations and more defense compounds as a result of higher atmospheric $CO_2$ concentrations. In particular, trees may become more susceptible to bark beetles, aphids and mature-foliage eaters. Insects that feed on newly emerging foliage may be adversely affected if the synchrony between egg hatch and bud emergence is disrupted.

Some fungal diseases of crops, native plants and animals could become more prevalent following mild winters. The frequency and distribution of bird diseases could change (e.g. duck virus enteritis, avian tuberculosis and botulism) as could mammal diseases and parasites. Also, the number and diversity of viruses could increase, depending on changes in agriculture and the activity and spread of vectors.

### Productivity

An increase in temperature of 1–3°C in the UK is likely to increase total plant productivity, and thereby increase some invertebrate and other animal populations (provided rainfall does not decrease, and where mineral nutrients are available), because of the lengthening of the growing season, especially in spring when solar radiation receipts are high. The argument that an increase in respiratory losses of carbon will exceed the increase in assimilation is likely to be true only for trees. The projected increase in $CO_2$ concentrations will also increase the productivity of $C_3$ plants (see Box 1, Direct Effects of $CO_2$ on Plants). However, species will differ in their growth and development responses to high $CO_2$ concentrations and these

**Box 1: Direct Effects of Carbon Dioxide on Plants.**

Almost all plant species currently growing in the UK use the $C_3$ biochemical pathway for photosynthesis. There are two general effects of enhanced $CO_2$ concentration on $C_3$ plants: first, growth is increased because rates of photosynthesis increase and rates of photo-respiration decrease; second, increased $CO_2$ causes stomatal pores on leaf surfaces to close partially and this reduces water loss per unit of $CO_2$ assimilated (i.e. it improves the 'water use efficiency'). There is some evidence that plants grown at high $CO_2$ concentrations for long periods partially acclimate to those concentrations and produce less of the enzyme (rubisco) responsible for assimilating $CO_2$ and fewer stomata. Such acclimations would lessen the impact of high $CO_2$ on photosynthesis and reinforce its impact on water use efficiency. There is also evidence that high $CO_2$ can alter plant development and increase the ratio of root to shoot mass.

The total water use of plants growing in high $CO_2$ may not be reduced, although the water use efficiency of individual leaves will be, because plants in high $CO_2$ tend to produce larger leaf areas from which the water evaporates. It is unlikely that increased $CO_2$ will change water use by vegetation significantly in the absence of changes in other variables that influence evaporative demand.

Plants such as maize, which use the $C_4$ pathway for photosynthesis, would not increase their $CO_2$ uptake in high $CO_2$ concentrations, but they are more efficient than $C_3$ plants at assimilating $CO_2$ at higher temperatures. Their stomata respond to increased $CO_2$ in a similar manner to $C_3$ plants, and so their water use efficiency would improve.

There is much variation among species in response to high $CO_2$ concentrations and their responses depend on other environmental variables (e.g. light, nutrition, temperature). However, there is a concensus that increasing $CO_2$ will increase the productivity of most, but not all, species in the field and will tend to mitigate the deleterious effects of water stress, low light and poor nutrition.

---

differences may be an important cause of changes in the competitive balance among species.

*Landscape changes*

The migrations, losses and variable responses of species mentioned above will change many of the landscapes with which we are familiar.

- Montane plant communities may be lost.
- Heaths may be subject to more frequent burning if it became hotter and drier, and would be invaded and replaced by other communities.
- Fenlands may dry out, especially if the current level of water abstraction from aquifers was increased.
- Blanket peatlands could change greatly in character if there were both warming and decreased rainfall.
- Salt marshes and brackish water habitats (estuaries, sea lochs and grazing marsh ditches) may change with sea level rise (see below).
- Permanent pastures may change because large differences occur between species in the optimum temperature for growth.

Other possibilities could be given, but confident predictions must await research.

### 4.2.2 Effects of sea level rise

Approximately 10% of the notified nature reserves (NNRs and SSSIs) of the UK occur near sea level on the coast.

A 20–30 cm increase in sea level will affect the lowest areas of mudflats and may significantly affect areas of salt marshes, particularly if landward expansion was prevented by the presence of sea walls.

The invertebrate fauna of intertidal flats is likely to become poorer and less diverse. The productivity of the surviving species may be reduced by high loads of suspended sediments in the inshore waters.

These changes in mudflat extent and in the abundance of invertebrate fauna would greatly reduce the numbers of many species of birds that roost, feed or breed on the UK coasts. For instance, the UK coasts are the wintering grounds for over half of Europe's waders, and 60% of the UK redshank nest in salt marshes.

## 4.3 UNCERTAINTIES AND UNKNOWNS

All reviews of the impacts of climate change on the UK flora, fauna and landscape stress that the two major unknowns are, (i) the future patterns of land use, and (ii) changes in the return periods of extreme weather events, especially droughts. The impacts of climate change on natural systems will remain uncertain as long as the climate of the future is ill-defined. Nevertheless, much can be done to explore the complex relationships between organisms and climate to address the major uncertainties that are listed as questions below.

- What changes in land use will occur in response to climate change?

- Which of the warmth-loving species of plants and animals that occur in southern Britain are most likely to spread northwards following climatic warming of 0.5, 1.0 and 2.0°C, and what is the likely rate of migration? What are the relationships between climate and the distribution of plants, butterflies, other invertebrates, birds and mammals? Can species be identified that will adapt to climate change rather than migrate?

- Which exotic species are likely to invade or expand and become hazards as weeds, pests, diseases or viruses? What are the relationships between climate and the distribution and outbreak of insect pests, weeds, diseases and viruses?

- What magnitude of climate change would bring about extinctions, and which of the sensitive species or communities listed above (losses and extinctions) will be affected first?

- What are the critical climatic variables (temperature, soil water deficits, vapour pressure deficits etc.) that may initiate the decline of different types of forests, in different soils, in the UK?

- What is the magnitude and significance of the various effects of $CO_2$ fertilisation in field conditions?

## 4.4 PRINCIPAL IMPLICATIONS

The following points summarise the key issues raised in the foregoing discussion on the estimated effects of climate change on the UK natural biota.

- Land-use changes brought about by socio-economic decisions or climate change could influence the natural biota as much as climate change itself.

- The natural biota are sensitive to change in almost all aspects of climate and weather. The variables to which they will respond most may be ranked: extreme events or seasons > soil water deficits > mean temperatures $>/=$ $CO_2$ concentration increase from 350 ppmv to 530 ppmv.

- There will be significant migration of species northwards (and to higher elevations), at a potential rate of 100 km/decade. Invertebrates, birds and 'weedy' plant species will move first.

- There will be an increased probability of invasion and spread of alien and potentially harmful weeds, pests, diseases and viruses.

- Many native species and communities will retreat and some may be lost to the UK. The losses of both plants and animals will occur particularly within (i) montane communities, (ii) coastal communities, (iii) confined, island habitats, and (iv) wetlands and peatlands.

- The natural flora is likely to be impoverished by this loss of many endangered species, but the invertebrate and avifauna may be enriched in much of the UK.

- Natural wetland areas could shrink and many could disappear.

## 4.5 RESEARCH AND POLICY NEEDS

### 4.5.1 Future research effort

*Framework*

Relationships between the UK flora and fauna and climatic factors need to be researched in a way that enables *predictions* to be made about the effects of given climate scenarios. Given the complexity of biological systems, it is suggested that the central activity will often be the construction of predictive models of various kinds (statistical, process-based, dynamic etc.). For instance, models are needed to predict potential ('equilibrium') and actual plant and animal distributions, insect population dynamics, forest-climate relationships, and the responses of vegetation in the field to elevated $CO_2$ concentrations.

These models should be supported with data and information from two sources as shown in Figure 4.2:

(i) Experiments conducted in the field and in controlled environments, which examine the responses of individual species or processes to environmental variables, following what may be called a 'bottom-up' approach.

(ii) Analyses of historic and current field data, including the many long-term data sets referred to in Section 4.1, to establish general relationships between, for instance, the ranges and sizes of plant and animal populations and climatic variables, following a 'top-down' approach.

### Responses to temperature and rainfall

The priorities for research are to address the general questions posed above concerning *land use, migration, invasion, extinction* and *forest health.* Each of these topics will require one or several of the research frameworks and predictive models shown in Figure 4.2.

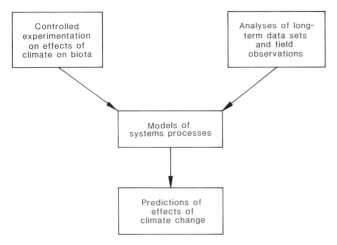

Figure 4.2 A simple framework for research on the impacts of climate change on flora and fauna.

### Direct effect of $CO_2$

The direct impacts of $CO_2$ on plants needs separate attention to determine, (i) species differences in growth responses to enhanced $CO_2$, including long-term experiments on trees, (ii) interactions between enhanced $CO_2$, temperature, water and nutrient supply, (iii) effects on plant development and key events in the reproductive cycle, (iv) effects on key communities of species, and (v) effects on absolute water loss as well as water use efficiency.

### 4.5.2 Policy recommendations

Policy issues need to be addressed concerning conservation and the risk of invasion or spread of harmful organisms. The policy issues concerning UK forests and woodlands are dealt with in Section 5.

### Conservation

(i)   The nature conservation strategy for the UK will need to be rethought. An increase in temperature of 1°C would significantly alter the species composition in over half of the statutory protected areas (NNRs, SSSIs, and other areas). With temperatures increasing by 0.3°C per decade, a dynamic nature conservation strategy will be needed to take account of continuous changes in species composition in protected areas. Management policies aimed at preserving existing communities within reserves will need to be reviewed. Existing reserves may provide sites where new communities and migrating species can be conserved, and the conservation values of existing sites will become different from those for which they were originally notified.

(ii)  A series of wildlife corridors and/or 'stepping stones' outside the NNRs and SSSIs should be developed to facilitate migration. In general, more consideration needs to be given to the habitat value of managed land that separates the reserves.

(iii) Questions of physically translocating species northwards, or of actively assisting migration, need to be addressed, with the aim of preventing extinctions and of creating new communities and wildlife reserves.

(iv)  Special consideration needs to be given to the creation of new coastal reserves near soft coasts, anticipating the disappearance of existing reserves.

### Pests

Regulations which restrict the movement of animals and living plant material into the UK may need to be strengthened and more rigorously enforced. A revised list of potential pests, diseases and weeds needs to be compiled, and provision made for coping with pest outbreaks. A policy may be needed on pesticide use to prevent large amounts of chemicals being used to control insects, diseases and weeds, rather than adapting to change.

# Agriculture, Horticulture, Aquaculture and Forestry

# 5

**SUMMARY**

- Experiments in controlled environments suggest that an increase of atmospheric $CO_2$ from the current value of 350 ppmv to about 450 ppmv by 2030 could increase the productivity of crops by 5–15%. However, it is uncertain whether this increase will occur in the field.

- With increased $CO_2$ levels, the amount of water used per unit of plant growth decreases but, because plants growing in high $CO_2$ are larger than those in ambient air, it is not clear whether the *total* water requirement of crops and forests would decrease.

- In general, even if there were sufficient soil water, higher temperatures would decrease the yields of cereal crops (such as wheat) although the yield of crops such as potatoes, sugar beet and forest trees would tend to increase. The introduction of new cultivars better suited to the changed climate could optimise these responses.

- The length of the growing season for grasses and trees would increase by about 15 days per degree C increase in mean temperature. This increase could improve the viability of grassland, animal production and forestry in the uplands. Higher temperatures would also create opportunities to convert currently unproductive upland moor and heath to grass pasture, but this would require considerable inputs of lime and fertilizer.

- More than 20% (by value) of UK horticultural crops are grown unprotected by glasshouses. A warmer climate would offer opportunities for raising high value over-wintered crops and for shortening the growth period of spring-sown crops. Glasshouse heating costs would be reduced, but any increase in winter cloudiness could decrease the production potential in glasshouses.

- Problems such as Colorado beetle on potatoes and rhizomania on sugar beet, currently thought to be limited by temperature, could become more important in the future, as could the risk of pest outbreaks on forests.

- If temperatures increase, there would be opportunities to introduce into the UK new tree species and crops that are currently grown in warmer climates. Maize and sunflower might be grown for their seed/grain yield as well as for fodder over much of the arable area of the UK if temperatures rose by about 1.0°–1.5°C from present values. Workability of land and weather extremes (wind, frost) might constrain their introduction.

- Some forests or woodlands growing on poor, dry soils may become unhealthy or die following a succession of dry summers, despite longer growing seasons and $CO_2$ fertilization.

- Changes in climate might have large effects on soils. Land use policy may need to be reviewed in the light of this, with regard to the changing potential of soils for crops, trees and other plant species. Guidelines for decisions on the protection of certain soils and land may need to be developed. Monitoring of land and soil salinity should be incorporated into soil conservation policy.

- An increase in the frequency of long hot summers would be detrimental to the trout and salmon farming industries, both directly by limiting water availability, and indirectly through increased incidences of low oxygen concentrations in water and of disease.

## 5.1 INTRODUCTION AND BACKGROUND

Although the UK is comparatively only a small country, it supports a wide range of food production and forestry activities, principally as a consequence of the large variation in climate over the country. Broadly speaking, the variations are in mean temperature (decreasing by about 3°C from south to north), rainfall (decreasing by at least 50% from west to east) and radiant energy (decreasing by about 40% from south west to north east). The range of climates in which UK agriculture is practised already covers part of the range of changes in climate suggested for the future (see Section 2). There is no UK analogue for the warming that is projected for southern Britain, but experience from warmer parts of Europe may provide useful comparisons. Mild winters and hot summers in the UK provide useful indications of the short-term impacts of climate change (Unsworth *et al.,* 1989), but it must be emphasised that all projections of future crop productivity in the UK and elsewhere are limited because we do not know the impact of changed $CO_2$ and the interaction of $CO_2$ with other environmental factors (Morison, 1988; Goudriaan and Unsworth, 1990).

There can be little doubt that UK agriculture can adapt to climate change if it is gradual, as assumed here; the more relevant needs are to identify regions, crops and production systems likely to be most sensitive to change, and to establish the sensitivity of fundamental processes in plant-soil relationships.

The UK agricultural industry also responds rapidly to external influences such as EC policy, world trade opportunities, etc., and this makes it impossible to fully assess responses to climatic impacts in isolation. For example, future climates might favour enlarging the area for the production of sugar beet in the UK, but, as a member of the EC, the economic prospects for the crop would also depend on the situation elsewhere in the EC and on the influence of climate change on sugar cane production in the tropics. Similarly, production of some horticultural crops, including the economics of glasshouse production, is likely to be very sensitive to the conditions for horticulture in other producing countries.

As the climate changes over time, it is unlikely that large scale sudden switches in agricultural and horticultural enterprises will take place. Forestry practice, where crop rotation times are several decades, is even less likely to respond rapidly to climate change. On the other hand aquaculture is rather vulnerable to hot dry summers. Farmers are most likely to continue with existing enterprises for which they have experience and capital investment, probably modifying their management schemes by introducing new cultivars of established crops. In parallel, some of the more enterprising farmers will experiment with small areas of alternative enterprises. Changed frequencies of extreme events such as droughts, gales and floods may add step changes, accelerating this steady adaptation. It is therefore useful to review first the sensitivity of existing enterprises in the UK to climate change and then to comment on the scope for alternative agricultural activities.

Almost three-quarters of all agricultural land in the UK (about 13 million hectares) is covered by grassland and grazing land. This land use completely dominates the landscape of the north and west of the country. Most of the remaining arable land, predominantly in the south and east, is planted with five major crops: wheat, barley, oil-seed rape, sugar beet and potatoes. Horticultural crops of vegetables and fruit make up the remainder. Over 85% of the area of arable crops is sown in the autumn when the soil is best suited for tillage.

## 5.2 ESTIMATED EFFECTS OF CLIMATE CHANGE AND SEA LEVEL RISE

### 5.2.1 Effects of climate change

*Principles*

In most crop systems, apart from rough grazing, the land is fertilized to minimise limitations from mineral nutrition. Water stress can be a limiting factor in the south and east of the UK, where irrigation may be used on high value crops which are drought sensitive. There is a strong correlation between the amount of light intercepted by a crop over the season and the yield, and consequently annual crops are managed to attempt to develop leaf cover as soon as possible and to ensure its duration for as long as possible, within the limits imposed by the need to harvest when weather and soil conditions are likely to be most favourable.

Temperature is the strongest factor in the UK influencing the development and canopy expansion of annual crops, and it is also the main constraint on the productivity of sown perennial grassland, but growth of indigenous grassland is often limited by nutritional deficiency. Frost-sensitive crops, such as potatoes and tender horticultural crops, cannot be sown until risks of spring frosts are low, and in some regions their growing season is cut short by the first frosts of autumn.

In relation to temperature response, it is important to distinguish between determinate crops such as cereals, and indeterminate crops such as grass and sugar beet; in the former, warmer temperatures shorten the duration of the necessary successive and finite stages of growth (e.g. leaf emergence, flowering, grain filling, ripening), and consequently shorten the duration of the season over which the crop intercepts light. Consequently, increased temperatures decrease the yields of determinate crops (Monteith, 1981; Monteith and Scott, 1982; Squire and Unsworth, 1988; Goudriaan and Unsworth, 1990).

In contrast, indeterminate crops continue to produce leaves and to grow for as long as the temperature remains suitable, and so yields of these crops tend to increase as temperature rises.

Rainfall influences crop production in two ways. First, excessive soil moisture makes cultivation difficult and damaging, leaches nutrients, inhibits microbial activity, and limits plant growth because of water-logged root systems (see Section 3). Conversely, soil water deficits in summer can restrict canopy expansion and productivity. If the air becomes drier on a regional scale, this can restrict the growth of some crops, irrespective of local soil water status (Monteith, 1981). For fundamental physiological reasons there are strong relationships between the rate of water use and the rate of crop growth; potentially this allows predictive modelling of responses to altered rainfall, but the relationship depends on soil and atmospheric properties in ways that are not yet fully understood (Squire and Unsworth, 1988).

UK crops and trees almost all use the $C_3$ pathway of photosynthesis to absorb carbon dioxide from the atmosphere and convert it to plant material (see Section 4, Box 1). All $C_3$ plants increase their photosynthetic rates when $CO_2$ increases. Based mainly on results from controlled environment experiments, an increase of $CO_2$ from the current value of 350 ppmv to about 450 ppmv by 2030 would potentially increase the growth rate of annual crops by about 5–15%. Preliminary results from trees and natural ecosystems growing in doubled $CO_2$ in chambers in the field show much greater increases in growth (150% in two years) (Long, 1990). It is not clear whether this results entirely from the cumulative effect of $CO_2$ on perennial species, or whether previous laboratory and controlled environment studies have involved unidentified limiting factors for growth. It is therefore uncertain whether reported results for crop species would apply in the field. Increased carbon dioxide also causes partial closure of stomatal pores, making plants use less water per unit of growth, although the total water use per plant may not decrease.

Crops such as grain maize, which might be more widely grown in warmer climates, use a different biochemical pathway to fix $CO_2$ (the $C_4$ pathway, Section 4, Box 1) and their growth does not benefit so much from higher $CO_2$ concentrations, although their water use efficiency is improved. In high $CO_2$ concentrations, some laboratory studies have shown that root growth of $C_3$ and $C_4$ plants increases more than shoot growth; even so, the nutrient value (e.g. nitrogen content) of above-ground parts tends to be lower than in plants growing in normal $CO_2$ (Kimball et al., 1986).

*Cereals*

There is a good understanding of how UK cereals respond to environmental factors in the field for all factors except $CO_2$. Analysis of past weather records supports the view that the main factor influencing winter wheat yields on heavy soils in the UK is temperature; in the East Midlands of England the change in yield is about minus 6% per 1°C increase. Inadequate rainfall contributes to variation in yields, especially with winter barley and spring cereals on light soils (Monteith, 1981).

On the basis of knowledge of environmental responses, several mechanistic computer models of cereal growth have been developed (e.g. see reviews in Day and Atkin, 1985). In particular a model of winter wheat developed at the Institute of Arable Crops Research generally predicts potential growth and yield well, even in conditions several degrees warmer than the UK averages (Porter, 1984). Squire and Unsworth (1988) modified the model to predict growth in 680 ppmv $CO_2$ and in increased temperature, assuming adequate water and nutrition. They showed (Figure 5.1) that, for this determinate crop, grain yield increased by about 27% when $CO_2$ concentration was doubled from 340 ppmv to 680 ppmv. But as temperature increased, the grain yield decreased, until, with a seasonal temperature 4°C above the control case, the benefits of doubled $CO_2$ on yield were lost, although there continued to be large increases in straw production. Porter (1989) combined the model with a soil-water and nitrogen model, showing that if nitrogen was applied according to current practice, the combination of high $CO_2$ with warmer, wetter winters resulted in substantial nitrogen leaching to ground water; this would also accelerate soil acidification and increase the need for liming. Mitchell (1989) showed that if warmer summers were cloudier, a 5–10% decrease in solar radiation would cancel the benefits of doubled $CO_2$. In all of these examples it must be emphasised that the adjustment of crop photosynthesis to higher $CO_2$ concentrations was based on the results of controlled environment studies which may not be valid in the field.

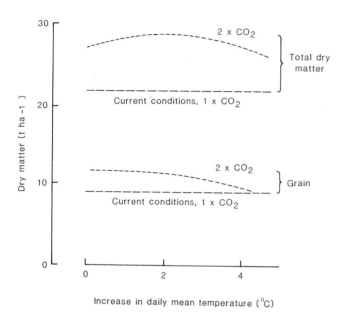

Figure 5.1 Calculations using the AFRC Wheat Model (Porter, 1984) modified for $CO_2$ responses by Squire and Unsworth (1988) to illustrate the potential production of winter wheat. The dotted curves are the grain and total dry matter yield in typical current conditions; the full curves show the effects of doubled $CO_2$ and increases in daily mean temperature of up to 4°C. (Source: Squire and Unsworth, 1988).

### Potatoes and Sugar Beet

The growth of these indeterminate crops would be likely to benefit from increases in temperature and $CO_2$ but effects of $CO_2$ on potential yield cannot be estimated precisely for these crops. Squire and Unsworth (1988) estimated on the basis of calculations by B. Marshall (Scottish Crop Research Institute) that potential yields of potatoes might increase by 15–25% with a 2°C increase in temperature alone, and by about a further 5% if, in addition, $CO_2$ were doubled. Both crops are rather sensitive to drought: 25% of the UK area is irrigated, including an important part of the main crop potatoes. With warmer and drier summers it is possible that drought in eastern Britain would limit the potential for these crops. Potatoes might be grown further north and west in areas of higher rainfall, but the economics of sugar beet require it to be close to the processing factories, most of which are in the east of England. In warmer climates, problems of virus diseases in potatoes and sugar beet would probably increase, e.g. there might no longer be virus-free areas for seed potato production.

### Grass

Most grassland plants in the UK have $C_3$ physiology and indeterminate growth, and would *probably* respond to a 30% increase in $CO_2$ concentration and associated increases in temperature by a 5–15%

increase in photosynthetic rate, and improved water use efficiency, but with decreased quality and digestibility of the plant material for the ruminant. However, by far the most important effect of increased temperature would be to increase the length of the growing season for the perennial crop, by increasing the period of grassland growth during the winter. Figure 5.2 shows the substantial growth

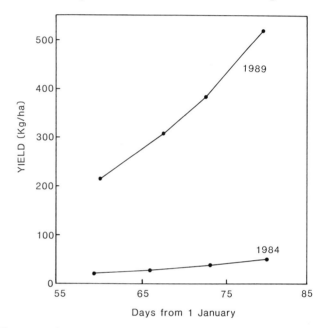

Figure 5.2 Grass growth and yield over winter 1983/84 (a typical year) and 1988/89 (very mild) at North Wyke, Devon. Plots were cut at the beginning of November, and the subplots were harvested at various dates from late February to assess the accumulated growth since November. (Source: Unsworth *et al.,* 1989; data courtesy of Institute of Grassland and Environmental Research).

of grass at a site in Devon during the mild winter of 1988–89, when local temperatures were about 1.5°C higher than normal. A 2°C average increase in temperature in spring and autumn would extend the growing season in most grassland areas of the UK by about 30 days. Since such areas of upland and hill grassland in the UK are currently marginal in terms of economic return because of altitude or terrain, a small increase in temperature could have significant effects on the viability of grassland in these areas. Much western UK grassland on heavy soils cannot now be used effectively in early spring, because of its water-logged state. Wetter winters might exacerbate this situation, so that the benefits of improved winter growth could not be achieved, but soil conditions depend on the balance between rainfall, drainage and evaporation, and cannot be predicted on the basis of current information on projected climate changes.

Increases in temperature could allow the conversion of unproductive heathland into grass/clover-dominated pasture if the nutritional status were satisfactory. While such plants could be encouraged to grow at higher altitudes under a warmer climate, this would occur only through considerable inputs of lime and fertilizers to increase soil pH. There are considerable uncertainties about how warmer temperatures might affect soil microbiology and consequently nutrient mobilization in these upland soils.

## Horticultural crops

In terms of value, more than 70% of UK horticultural crops are grown unprotected; tomatoes and mushrooms are the main protected crops. The timing of harvest and quality of produce are very important variables in the economics of horticultural production. One of the greater benefits of climatic warming could be the improved opportunities for growing overwintered crops for harvest in early spring, especially if the maritime climate of the UK gave a lower risk of frosts than in nearby continental countries. Other benefits of warmer temperatures include quicker turnaround from sowing to harvest, improved quality of some crops (e.g. skin colour in onions), and reduced heating costs in glasshouses. However any increases in winter cloudiness, i.e. lower light levels, would partly offset savings in heating in the glasshouse industry and would affect the quality of some field-grown horticultural crops. Other potential negative effects of the climate projections described in Section 2 include alterations in the timing and degree of vernalisation of crops such as carrots (increasing risks of bolting), reductions in winter chilling necessary for some top fruit such as apples and pears, increased risks of pests and virus disease, and problems of seed-bed preparation (poor germination) if soils tended to be drier.

## Animal Production

Increases in temperature would have an insignificant direct effect on outdoor farm animals in the UK. There are currently important economic losses of newborn lambs in the uplands because of wind-chill, and these losses would reduce if the incidence of cold, wet, windy conditions decreased, but warmer temperatures alone would not help significantly. With the increased likelihood of hot summers, there could be difficulties in providing adequate ventilation for housed animals such as poultry. In hot summers such as 1990 there is evidence that some animals in the field suffer heat stress, but losses in production have not been quantified. It has also been suggested that heat stress affects the fertility of cattle, sheep and pigs.

Grassland provides 60–90% of the energy requirements of all our ruminant animals. Animals cannot be acquired or disposed of abruptly without financial loss, and viable farming enterprises are built around a conservative estimate of carrying capacity, largely defined by winter shortages of foodstuffs. For this reason, grassland is often under-utilized in the growing season. If warmer climates reduced the winter 'gap' in grass growth, this would reduce the need for conserved feed, or bought-in supplements, and could have disproportionately large effects on potential animal carrying capacity, the economic viability of the grassland farm, and the profit to the farmer. This has implications for the viability of investing in land improvement schemes in upland and marginal areas. Mild winters alter the lifecycles of pests and parasites of livestock, and, depending on management practices, may alter the incidence of infection (Unsworth, et al., 1989).

## Freshwater Aquaculture

Temperature increase and in particular an increase in the frequency of long hot summers would have a significant effect on the aquaculture industry. Salmonids, the main commercial crop, would grow faster for much of the year, but the summer period (when dissolved oxygen levels are too low for feeding) would be extended and this, coupled with drought effects on water availabililty, would render much of the English trout industry non-viable.

It may also reduce the range of viability of salmon farming to the North of Scotland. This would significantly affect both freshwater and marine farmers, and have severe economic effects on the fragile economies of the rural Western Highlands.

One of the first manifestations of climate change would be an increase in epizootic disease problems, which always follow high temperatures and low dissolved oxygen levels. This in turn could affect usage of antibiotics and pesticides with both economic and potential residue problems unless more rational control methods were developed.

The rise in temperature may however provide opportunities for culture of species currently not viable in the UK. It also introduces the possibility of survival in the wild of exotic species cultured or imported for the aquarium trade, and this would be detrimental to both farmed and native wild species.

## Forestry

UK forests are highly vulnerable to any increase in mean windspeeds or in the frequency of great storms. The risk of windthrow is always a major constraint on commercial forestry in the uplands. Any increase in windspeeds will place a greater area of forest in high 'windthrow hazard classes', where thinning is restricted and the trees must be felled before they reach an optimum height for sawlog production. Major forest fires, which are currently rare in the UK, would become more likely in southern Britain if there were an increase in the frequency of hot, dry summers such as 1976 and 1990.

In the absence of wind damage, fires and pest outbreaks, forest growth is likely to be enhanced by a temperature rise of 1–2°C, especially in the north and the uplands, and where a warming also enhances soil mineralisation and the release of nutrients (Cannell, et al., 1989). Conifers are already grown at a range of locations in the UK differing in mean annual temperature by at least 2°C, and trees growing at the warmest locations generally have the highest productivity, provided there is adequate rainfall.

However, Sitka spruce is adapted to humid conditions and would be adversely affected if an increased frequency of hot, dry summers with low humidity occurred. The health of trees may be affected for several years after hot dry summers. This may be the result of a combination of stresses, including high concentrations of air pollutants such as photochemical ozone. If climate change resulted in a succession of dry summers, these could trigger the onset of irreversible changes in those UK forests and woodlands where tree condition and growth are already adversely affected by soil water deficits, poor nutrition and other stresses.

*Pests, Disease and Weeds*

The principal concerns raised by climate change are:

● Will the spectrum of pests and diseases found in the UK alter?

● Which, if any, diseases are likely to become more serious?

● Will pest and disease control be more difficult?

Plant diseases are caused by viruses, bacteria and fungi. Virus diseases depend on vectors for spread. Bacterial and fungal diseases of plants are affected by climate both directly, through the effect of temperature and moisture on their life cycle and on the development of the disease they cause, and indirectly through changes in the susceptibility or resistance of the host plant. Similar climatic factors affect animal diseases transmitted by airborne viruses and bacteria, and pests and parasites in soils and food.

Experience in the mild winter of 1988/89, which was 2.5–3°C warmer than average in many parts of Britain, indicated that rust and mildew diseases of cereals were widespread. This was partly because the mild conditions favoured the growth and spread of the pathogens, but also because weeds and

volunteer crop plants growing throughout the winter created a 'green bridge' on which the diseases could thrive and then affect spring-sown crops. The green bridge is also important for the survival of insect pests such as aphids, which are the principal vectors of plant virus diseases. There are strong correlations between the timing and severity of applied outbreaks and the mean winter temperature (Fig. 5.3). In 1989 and 1990, after the mild winters, aphid infestations were much earlier and larger than normal (Harrington, 1989; Unsworth, et al., 1989).

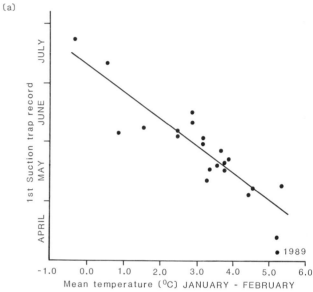

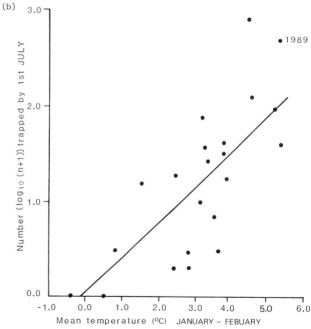

Figure 5.3 Relationships at Harpenden between the incidence of peach-potato aphid (a vector of potato virus disease) and winter temperatures: a) date of first detection in suction traps, b) number of aphids trapped by 1 July each year (note that axis is logarithmic). Observations after the mild winter of 1988/89 are specifically labelled. (Source: Harrington, 1989).

Weeds are, by definition, opportunistic species, particularly well adapted to take advantage of mild winter conditions. The current practice of herbicide application on overwintering crops makes use of the extra stress of frost to kill herbicide-treated weeds. In milder winters this would not be successful, and new approaches to weed control would be necessary.

It is likely that with climatic warming and the introduction of new crop species, some new pests and diseases will become established, whilst other existing problems will become less serious. Potentially serious problems such as Colorado beetle on potatoes, and rhizomania on sugar beet, currently both thought to be limited by temperature, could become much more widespread. As an example, in the mild autumn and winter of 1989–90, populations of the European ground beetle (an indigenous species in southern Britain but not usually a problem) built up on south-facing fields in Sussex and caused widespread damage to cereal crops in spring.

The coordination of pest (or disease) and plant development along with controlling factors such as predators, are important factors determining the scale of attack. Consequently the impacts of climate change on the crop/pest/predator complex need to be assessed.

### New Crops

As temperatures increase there will be opportunities to introduce crops into Britain that are currently grown in warmer climates. The most likely introductions are crops already widespread on the continent, for example grain maize and sunflower. Parry and his colleagues have shown that in terms of thermal time, both crops could be grown for their seed/grain yield as well as for fodder over much of the arable area of the UK if temperatures rose by about 1–1.5°C from present values (Parry, et al., 1989). Their introduction is more likely to be limited by the frequency of extremes such as frosts, and by the need to fit them into current agricultural practice especially the timing of harvests, land preparation and sowing. The economic viability of these and other new crops would depend critically on changes in the climatic conditions in the regions where they are grown at present.

If temperatures increased by more than 2°C, it would be advisable and perhaps imperative to restock much of the existing commercial forest with different provenances or species. Native Scots pine and some native broadleaved species may be ill-adapted to withstand high temperatures, especially in an absence of winter chilling. There would be an opportunity to plant more broadleaved trees (native or exotic) over a wider area of the UK, where soil permitted, and many more exotic broadleaved tree species could be successfully introduced (Cannell et al., 1989).

### 5.2.2 Effects of sea level rise

Effects of sea level rise would largely be through effects on soil, either by reduced drainage, salination or inundation (see Section 3). About 720,000 ha of agricultural land lies within the area bounded by the 5m above-sea-level contour, and much of this is of very high quality. About 8% of the 8.2 million hectares of grades 1–3 agricultural land lie below the 5m contour, but 198,000 ha of this fraction are grade 1 land (i.e. 57% of all grade 1 land in England and Wales) (Whittle, 1990). The most damaging floods in the past century have been the river floods of 1947 and the tidal inundations of 1953. Areas important for agriculture and horticulture that are potentially vulnerable include parts of East Anglia and Lincolnshire, and the area around the Humber estuary. However, because most of these areas are protected by substantial sea-defences, they may be at less immediate risk of damage from flooding than low-lying coastal and estuarine agricultural land elsewhere in the country.

### 5.3 UNCERTAINTIES AND UNKNOWNS

There are four major sources of uncertainty:

(i)  Information on possible future changes in climate in the UK is currently insufficient for prediction of the full range of impacts on agriculture. In particular the lack of quantitative guidance about winter and summer rainfall makes it impossible to know whether the land would be suitable for agricultural operations that could take advantage of a longer growing season. It is also uncertain whether, in the main growing periods, any increased potential for growth associated with $CO_2$ and temperature might be limited by water stress. Equally, lack of sound knowledge of the frequency of extreme events such as frosts and gales renders it impossible to assess the risks of introducing new sensitive crops as the mean climate warms up, or to assess the altered risks of weed damage to forests.

(ii)  There is a substantial base of knowledge of the ways in which many major annual and perennial crops and some forest trees respond to environmental factors such as light, temperature, water and nutrition, but there is little knowledge from studies in field conditions of the direct effects of carbon dioxide and of its interactions with these factors. The influence of changed climatic conditions on both the yield and nutritional properties of crop plants is uncertain; nutritional aspects are particularly important for the grazed grassland ecosystem.

(iii) The critical climatic parameters (temperature, soil water deficits, vapour pressure deficits, etc.) that may initiate the decline of different types of forests in different soils in the UK are not known, nor is the influence of air pollution.

(iv) The ecology of pests, pathogens and weeds in agricultural and forest systems is relatively poorly known at present, and cannot be predicted for future conditions. This knowledge is important for pest and weed management under the environmental constraints that are likely to be required in the future.

## 5.4 PRINCIPAL IMPLICATIONS

The most critical factor throughout the UK is likely to be temperature, both mean temperature and extremes. In the east and south of the country, decreases in rainfall could necessitate major alterations in the type of agriculture that is practised. Extremes of wind pose significant hazards for forests.

Although the increase of $CO_2$ will probably not accelerate, of all the factors concerning climate change, it is the most certain to change over the next few decades. Consequently it is important to establish the sensitivity of agricultural crops and forests to the major features of temperature, water and nutrition in relation to increased $CO_2$ concentration. We do not know at present how critical the sensitivity of crops and forests to $CO_2$ may be, and we cannot reliably identify the new opportunities for agriculture and forestry.

Impacts of changes of climate on other aspects of the environment and economy of the UK may have major implications for agriculture and forestry, for example:

● Soils – the physical structure, microbiology and nutrient properties of soils as influenced by climate are critical for agriculture, horticulture and forestry (Section 3).

● Water industry – implications of altered use of fertilizers and pesticides are likely to be extremely important for the water industry. The availability of water for agricultural irrigation schemes needs to be considered. Altered water use by forests growing in water catchments would be important for water resource planning (Section 7).

● Flora, fauna, and landscape – climate change is likely to have a major influence on land use capability for agriculture and for other activities (Section 4).

## 5.5 RESEARCH AND POLICY NEEDS

### 5.5.1 Future research effort

A number of topics have been identified where further research is necessary to establish sensitivities and understand processes and mechanisms:

(i) Studies of the responses of major crop species and cultivars to the interacting factors of temperature, water, mineral nutrition, and carbon dioxide. Some of this research needs to be done in conditions that simulate the field; other work is needed at the physiological and biochemical level to understand the mechanisms of response and interactions.

(ii) Studies of the effects of climate change on the stability of existing pasture plant associations and communities and of existing legume symbioses. A range of research activities is needed to understand the stability in terms of the physiology and ecology of the systems.

(iii) Investigations of effects of climate change on the microbiology of soil organisms and the consequences for the flux of nutrients from soil to agricultural plants.

(iv) Studies of the ecology of pests, pathogens and weeds in agricultural and forest systems, and of their control.

(v) Continued research on 'forest decline' to determine the tree species and forest areas that may be adversely affected by a rise in temperature, changes in rainfall and altered frequencies of air pollution episodes.

(vi) Research to identify the changes in tree species or provenances that may be required or permitted following climate change. Research is needed on the effect of a rise in temperature on seed production and natural regeneration of native and exotic tree species in the UK.

Each of the above research activities should be planned as an integrated programme of laboratory and field measurements and computer modelling. The aim should be to develop validated computer models that are capable of making useful predictions as improved knowledge of scenarios of climate change becomes available.

### 5.5.2 Policy recommendations

Three aspects of policy require attention:

(i) Land use planning – under changed climates there will be altered capabilities of using land for agriculture, recreation, conservation, etc. A national policy needs to be developed to consider how we should adapt to the transient changes in climate that are expected.

(ii) Water resource planning – the area of agriculture in the south and east of England that is sensitive to drought may increase. Consideration of future water requirements for agriculture and for other uses needs to proceed.

There also needs to be close consultation between agriculture and water authorities to assess the implications of research suggested above and to adapt policies so that water resources can be protected from contamination by pesticides and fertilizers.

(iii) Future forests and woodlands need to be better adapted to a changing and variable climate. Heterogeneous mixed species forests may be better buffered against climate change than monocultures. There is also potential for more short-rotation plantations on good quality land; such plantation can be replaced more rapidly than long rotation forests, and the cost of catastrophic damage to them is less.

# Coastal Regions

## 6

**SUMMARY**

- Increases in mean sea level would lead to an increased frequency of extremely high sea levels and coastal flooding. If there were also increases in storminess, storm surges and waves would further increase flooding probabilities.

- Flooding would result in damage to structures and short-term disruption of transport, manufacturing and the domestic sector. In addition, longer term damage to agricultural land, engineering structures such as buildings and coastal power stations, rail and road systems, would occur in some areas due to saline effects.

- A number of low-lying areas are particularly vulnerable to sea level rise: these include, for example, the coasts of East Anglia, Lancashire and the Yorkshire/Lincolnshire area, the Essex mudflats, the Sussex coastal towns, the Thames estuary, parts of the North Wales coast, the Clyde/Forth estuaries and Belfast Lough.

- An increase in sea level would be likely to reduce the efficiency of groundwater and sewage drainage in low-lying areas, with consequent need for more pumping.

- Groundwater provides about 30% of the water supply in the UK. A rise in the water table associated with higher sea levels may increase the salination of groundwater over a long period, with consequent effects on agriculture and water supply.

- There is the possibility of the inundation of significant areas of wetlands and saltmarshes found near the present High Water line, with the probability of concomitant inshore migration of coastal ecosystems.

- The success of adaptive responses to changes in coastal flooding frequencies will depend on the rate at which these changes develop. This applies both for natural systems such as salt marshes and for socioeconomic activities such as manufacturing, mining, transport and recreation.

## 6.1 INTRODUCTION AND BACKGROUND

There have been several estimates of sea level rises expected in the next hundred years or so. Present scientific analysis (Figure 2.9) suggests rises in global sea level of perhaps 20cm ± 10cm by 2030. The most reliable evidence for recent change comes from direct measurements of sea level (Pugh, 1987a), but these extend back little further than the early 19th century.

Relative land-sea level movements for the most reliable UK stations are given in Table 6.1. The general trends are consistent with the geological evidence for a gradual uplift of the north of Britain and a gradual subsidence in the south east, due to adjustments following glacier-load removal. Vertical land movements must be measured independently in order to obtain absolute sea level changes.

Several major studies of possible impacts of sea level rise have been published (US Department of Energy, 1985: National Research Council, 1987; Wind, 1987; Pugh, 1990; Stewart, *et al.*, 1990). Doornkamp (1990) is of particular interest for the United Kingdom. These studies all agree that rising sea level would have adverse effects, particularly coastal inundation and erosion, unless remedial measures are taken.

Whatever the implications of sea level rise for coastal flooding, probabilities of extreme events will vary considerably as the hypothetical example in Figure 6.1 shows; changes will depend on the local statistics. Revised statistics of sea level variability and extremes should be computed for each individual location as part of any design procedures. As a general rule the coasts of East Anglia and the Thames Estuary have relatively steep probability curves (Type B), so that an increase in sea level alone has a smaller effect on flood return periods than along the south coast of England where the probability curves have a more gentle slope (Type A) (Graff, 1981).

These different effects (which do not consider changes in wave statistics) assume that the shape of the probability curves remains the same, and that they are uniformly uplifted by the mean sea level changes outlined. This basic assumption may not be valid because changes in mean sea level can influence tidal and surge dynamics, and hence their statistics. An increase in mean sea level will also affect the response of the sea to wind forcing. Both surge and wave statistics will thus be affected because frictional losses of energy from waves reduce in deeper water, whereas the effects of winds in generating surges will theoretically reduce as water depths increase. However, changes in mean sea level may be a secondary factor compared with changes in the frequency and severity of storms which generate these extreme conditions. Wind statistics could assist in verifying trends in surge and wave statistics.

Table 6.1  Linear mean sea level trends relative to land for selected UK ports

|  |  | Direct Analysis | | With Atmospheric Correction | | Estimated Vertical Land Movement |
|---|---|---|---|---|---|---|
|  | Data Span | Trend | SD | Trend | SD | (see text) |
| Newlyn | 1916–1982 | 1.72 | 0.16 | 1.78 | 0.11 | −0.3 |
| Devonport | 1962–1982 | 0.8 | 2.5 |  |  |  |
| Portsmouth | 1962–1982 | 5.0 | 0.5 |  |  | −3.5 |
| Sheerness/ Southend | 1916–1982 | 2.27 | 0.21 | 1.94 | 0.23 | −0.4 |
| Lowestoft | 1956–1982 | 0.3 | 0.7 |  |  | +1.2 |
| Immingham | 1961–1982 | 1.7 | 1.1 |  |  |  |
| North Shields | 1916–1982 | 2.57 | 0.22 | 2.61 | 0.24 | −1.1 |
| Douglas (IOM) | 1938–1977 | 0.26 | 0.67 |  |  | +1.2 |
| Aberdeen | 1916–1982 | 0.52 | 0.21 | 0.86 | 0.19 | +0.6 |
| Lerwick | 1958–1982 | −2.0 | 0.7 |  |  | +3.5 |

Note: All values in mm/year
Source: Based on Woodworth, 1987

Present knowledge cannot determine even the relative importance of the various factors contributing to the observed global sea level increase of 1.0 to 1.5mm per year (Figure 6.2). Although there is a high probability that sea levels worldwide will rise by more than 0.5m in the next 100 years, the uncertainties in these estimates are very large.

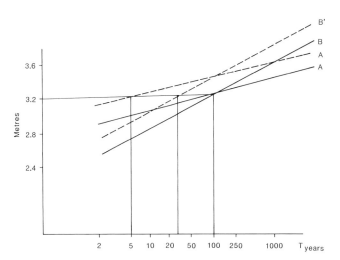

Figure 6.1 Showing how a change of 0.2m in mean sea level affects return periods for different probability curves. The 3.2m level (solid lines A, B) has a 'return period' of 100 years at both sites at present. However, for a 0.2m sea level rise the return period falls to 25 years and 5 years at B and A respectively (broken lines A', and B'). The examples are hypothetical, but B is characteristic of the east coast of England, and A is more representative of the south coast.

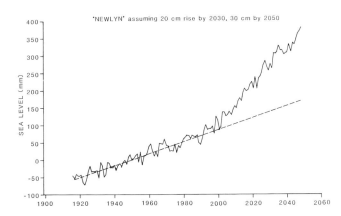

Figure 6.2 Newlyn sea levels with a superimposed trend and noise level showing the problem of identifying enhanced rates of rise against a natural oceanographic variability. Data to 1984 supplied by the Permanent Services for Mean Sea level; values after 1984 are the increases projected with the normal interannual variability superimposed. (Source: Woodworth, 1989).

## 6.2 ESTIMATED EFFECTS OF CLIMATE CHANGE

The frequency of extreme events will depend on weather patterns such as the tracking, frequency and intensity of storms. Changes will affect waves and surges, wetlands and dunes at the coastline, surface and groundwater salinity, water supply and drainage within the coastal zone. Natural and human adjustments to these different conditions will depend on the rate at which they take place as well as the total extent of the changes.

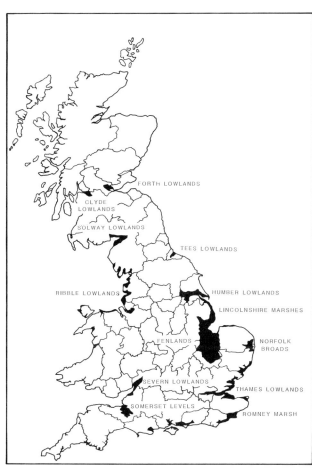

Figure 6.3 Areas of Britain where sea level increases could have a significant effect. The shaded areas indicate land which lies below 5m ODN. Projected sea level rises are at present much less than this. Local responses will depend on the rates of sea level change as well as the magnitude.

The principal effect of sea level rise will be an increased frequency of high sea levels and a greater probability of coastal flooding. At 1989 prices the cost of the 1953 east coast flooding has been estimated at £900 million. A 1000-year flood level incident on an unprotected London would inflict damage estimated at £10bn (1990 prices). Nationally an equivalent figure might be £100bn. Many of the areas at risk (Figure 6.3) are major conurbations or high grade agricultural land (Whittle, 1989). Several power stations (50%) are situated on low lying land (see Section 8.2.2.) as are many major manufacturing activities. For reasons of geography, many major communication systems, road and rail, are near the coast and will be increasingly at risk.

As sea level rises, groynes may have to be adjusted depending on the drift pattern where they are backed by dunes; they may eventually be outflanked at higher water levels unless the connection to the mainland is maintained. Consideration will have to be given to raising and strengthening existing defences. New defences may need to be designed to accommodate more severe conditions as more knowledge and confidence in the trends becomes available, over the design lifetime of a structure. As sea level rises, detailed local study and ongoing assessment is required into the viability of existing barriers, such as the Thames Barrier for the protection of London (see Box 2).

The consequences of not defending against sea level rise will vary with the nature of the coast. Erosion will generally occur although locally there could even be accretion of sediment. Dune erosion may be caused by changes in wind and sea level rises. Both could be serious problems, particularly where the dunes are part of a natural defence

system against flooding. Long term strengthening may be necessary. Options could include 'soft' technology such as the development of dune vegetation with a binding root structure (Whittle, 1989).

Coastal wetlands and saltmarshes which develop at levels near present High Water account for vast areas of land which would be particularly vulnerable (Boorman, et al., 1989). Much of this land is at the upper reaches of estuaries. Over the past several thousand years coastal wetlands and marshes have usually built up through sedimentation and peat formation at a rate comparable with the sea level increases. Within the UK, Boorman et al. (1989) estimate that for Essex, for example a 0.8m rise of sea level could lead to a reduction of at least 20% in the area of mudflats; any rise greater than about 1m could lead to a near total loss of marsh. The rates of sea level rise are very important in determining whether marsh ecosystems can adjust. Keeping land available for wetland and saltmarsh migration requires long-term planning. Ecologically, the saltmarshes serve as nurseries for marine and bird life. In addition they provide protection from storms and wave erosion to the hinterland and sea defences. (See also Section 4.2.2.)

Abstraction of water from the upper reaches of estuaries for domestic use and for irrigation would be affected by sea level rise. The effects can be compared with the salinity increases at times of drought and low river flow. Gradual salinity increases can be accommodated by upstream movement of brackish and freshwater species provided that space and water quality conditions are favourable. For human use water could be abstracted from higher levels in the river. Drainage of low-lying areas at low tide levels would become less effective; replacement of gravity drainage by pumped drainage will be necessary in some instances (Beran and Arnell, 1989).

Ground water provides about 30% of the water supply in the United Kingdom. Abstraction of groundwater in coastal areas at rates greater than natural replenishment by rainfall has already led to problems of brackish intrusion in the Merseyside and Humber areas (Edmunds, 1989). Higher sea levels would cause an upward movement of the water table and an increase in groundwater salinity over a long period. Coastal areas making use of chalk aquifers such as the Sussex coastal towns and the Yorkshire/Lincolnshire area may be particularly affected. However, in the overall context of climate change, changes in total and seasonal rainfall will be more significant for maintaining groundwater abstraction than sea level changes. UK cities (London, Liverpool, Birmingham, Nottingham) have experienced increases in water table levels of several metres in recent years because of reduced levels of abstraction for heavy industry, and because of mains leakages (Price and Reed, 1989); against this, effects of sea level rise on groundwater may be of secondary significance.

Commercial fish catch patterns are known to be very variable over years and decades, but the reasons for many of these changes are still unclear. Variations cannot be directly attributed to temperature changes in the seawater, and so predicting the effects of climate change on fisheries is very uncertain. If there are increases in water temperatures and in solar radiation, long term changes in both the quantity and composition of spring and summer plankton might occur. We might expect northern fish species such as cod, haddock and herring to be replaced by southern ones such as hake and sole. But even the direction of changes are unpredictable because the further we go from the physical processes towards the fish, the greater the uncertainties become.

A substantial rise of sea level would cause the nursery areas of young fish to move to new areas of shallow water, which could be more or less suitable. If formation of new areas of shallow water were prevented by coastal defences there would be a net adverse effect.

Within the constrained environment of fish farms there is increased susceptibility to solar ultraviolet radiation. Also, the problems of disease in aquaculture are recognised as being generally much greater during the summer months when there is a seasonal increase in water temperature. Thus, should sea temperatures increase there could be a greater susceptibility to disease.

However, in all these cases the scale of the probable changes is similar to that of naturally-occurring variability, to which adaptation has been possible in the past.

Indirect economic effects will include increased pressure for recreational use of beaches if the seas are warmer. However there could be reduced beach area as the sea level rise causes erosion. Socio-economic changes in the use of the coastal zone are difficult to anticipate, but cut across many areas covered in other sections of this report. Socio-economic effects of changes in fisheries are also probable. A recent example is the difficulty in re-establishing market demand for herring, and rebuilding the fish processing industry, following recovery of North Sea herring stocks.

## 6.3 UNCERTAINTIES AND UNKNOWNS

The problem of projected sea level rises may be addressed in a number of different ways. At one extreme vast resources might be expended on maintaining the present coastal position, and coastal activities, artificially. At the other extreme, a minimal response would be to plan an ordered withdrawal to higher land.

There seems to be a gradual change in public and institutional expectations for shoreline stability. Defence against erosion is no longer automatically expected at public expense. In the case of retreat, planning and timing are important elements and the wrong choice could be costly either way. Attempts to stabilise a shoreline in an area of high erosional tendency where the land was of low value would be too costly and may deprive other areas more deserving of a natural supply of material essential for coastal defence. On the other hand, if a natural system is only slightly out of balance and there is a large local tax base to fund remedial measures, then a decision to retreat could be wasteful.

Response to increased sea levels should be rational and progressive. However, even though the rise in sea level may be regular and designed for, the damage inflicted will be associated with extremes or clusters of notable events; major flooding events similar to the 1953 east coast inundations will trigger emotional and perhaps political responses at irregular intervals. In responding to these demands oceanographers and coastal engineers must be in a position to give rational advice based on well founded statistics on the probability of future flooding. For example, if a storm having a 500-year return period caused widespread damage it might not justify costly protective measures on cost/benefit arguments. However, a storm having a 40-year return period, which caused extensive damage, might well justify expensive protection. As a general rule we require sufficient understanding of the coastal environment to allow estimates of flood probabilities and their changes commensurate with the likely lifetime of the defences. Major sea defences and many coastal facilities, such as roads, bridges, docks, piers, coastal power stations using sea water for cooling (see Table 8.5), hotels and airports are designed and discounted over a lifetime of about 50 years. However, the associated infrastructure (roads, transmission lines, employee housing, etc.) may have longer-term stability. In terms of projected sea level rise this remains a relatively short period.

Nevertheless, uncertainties in the possible future rise in sea level might make it most effective to construct sea defences in such a way that they can cope with changes in storm surge patterns and sea level rise during their projected lifetime without excessive additional expense. It has been estimated (Dept. of Environment, 1988) that improvement of existing sea defences would cost in the order of £5-8 billion. In general, softer defences are less costly and have a shorter life than hard defences, and because they may have to be reconstructed more frequently, there will be opportunity to adapt to revised environmental design criteria as more knowledge becomes available and uncertainties are reduced.

There would be many advantages in protective systems which can be developed progressively in advance of sea level rise and for which the cost can be spread over a long period. These could include the development of protective vegetation to stabilise dunes and the encouragement of accretion and build up of wetlands and saltmarshes, if this is necessary to prevent their being overwhelmed by rising sea levels and storm events.

## 6.4  PRINCIPAL IMPLICATIONS

The probable major implications of projected sea level rises and increased storm surge activity can be summarised as follows:

- Increased probabilities of high sea levels and coastal flooding. This would lead to more frequent disruption to transport, manufacturing and the domestic environment.
- More persistent high sea levels and storms will affect the strength of coastal defences and other structures such as bridges.
- Reduced efficiency of groundwater and sewage drainage may occur, with consequences for pumping.
- The loss of wetlands and/or inshore migration of coastal ecosystems.
- The long-term saline intrusion of coastal aquifers.
- Changes in recreational coastal usage.

## 6.5  RESEARCH AND POLICY NEEDS

### 6.5.1  Future research effort

The first priority must be to improve the scientific understanding on which these predictions are based in order to reduce the uncertainties. Prediction should be viable over periods comparable with proposed capital developments.

The projected rise of sea level implies increases of only a few centimetres in the next 20 years. During this period we need to set in place comprehensive monitoring systems, both globally and nationally. We also need to establish a predictive capability linking global temperature increases and other aspects of climate change to sea level rise and we need to take account of the impacts of sea level rise and changes in storms causing waves and surges, in designing long-term coastal facilities. The following paragraphs itemise these research requirements in more detail.

The following long-term measuring and analysis activities are required to give the possibility of early warning:

*Globally*

i)  Development of the Global Sea Level Network (GLOSS) under the auspices of the Intergovernmental Oceanographic Commission (Pugh, 1987b), with special funding by developed nations; connection of selected gauges in this network to an absolute global reference system using the new geodetic technology.

ii)  Identification of causes of the interannual variability in the annual sea level measurements.

*For the UK*

i)  Maintaining and strengthening of the existing United Kingdom National Tide Gauge Network; these gauges should be connected into the absolute reference system.

ii)  Interpretation of data from these gauges to evaluate the effect of sea level rise and storms on flooding probabilities at specific sites, and trends in the statistics, including the interactions between tides and storm effects.

iii)  Evaluations of average High Water and Low Water levels for trends which may be different from those in mean sea level.

iv)  Local studies to determine changes of sea level in the recent geological past and consequent shoreline retreat or advance, related to local geology.

The impacts of changes, and possible responses, will vary considerably from place to place.

Even at this early stage it would be useful to initiate a series of case studies, as has been done for several areas in the United States, to identify the most significant potential impacts and possible responses in these regions (Figure 6.3). These could include extensions of recent Tyne-Tees and East Anglia studies, and studies of the Humber Estuary where the development of the Selby colliery complex is an interesting example of possible impacts of sea level change on mining. For policy and comparison purposes, these studies should apply similar techniques and standards to identify economic implications.

The design of new coastal facilities with a projected lifetime of several decades should take into account possible effects of sea level rise on the coastal regime as well as possible trends in extreme events; these studies could include analysis of different patterns of wave refraction and attack, and tidal and surge changes resulting from the different climate conditions.

Regional studies should consider the application of natural protection mechanisms, such as plant species which could stabilise a particular coast (Boorman, *et al*., 1989).

## 6.5.2. Policy recommendations

It is evident that there is no single best policy for adjustment to projected climate change impacts. It is also evident from other sections in this Report that each policy will have different direct and indirect economic implications: for example, a policy to build extensive sea walls or other protection would give a major boost to the quarrying industry but might adversely affect recreational facilities. Implications and possible responses need to be evaluated on a local basis, but these regions of study must not be too narrowly defined because coastal protection in regions of longshore sediment transport will inevitably cause erosional difficulties elsewhere. Economic benefits of protecting coastal facilities such as docks, roads and recreational beaches will also have regional scales, and it would be reasonable to expect the cost of their protection to be spread over a large hinterland.

Sea defences should be constructed in such a way that they can cope with changes in flood probabilities in their lifetime without excessive additional expense. It is necessary to develop a monitoring and prediction capability for coastal impacts of climate change.

A comprehensive research programme for identifying climate change impacts at the coastline and responding to such impacts will involve many separate scientific and engineering disciplines. The policy must be to develop a mechanism for predicting changing flooding probabilities, and a cost-effective strategy for adjusting.

If there is an adverse trend in storms and accelerated sea level rise in the 21st century then, for coastal regions, the associated sociological and political problems could require difficult decisions. But they will be more tractable if we can prepare for them in the next 20 years by establishing a sound understanding of the scientific and engineering implications.

# Water Industry

## 7

**SUMMARY**

- Wetter winters would benefit water resources in general, but warmer summers with longer growing seasons and increased evaporation would lead to greater pressures on water resources, especially in the south and east of the UK. Increased variability in rainfall, even in a slightly wetter climate, could lead to more droughts in any region of the UK.

- At present water demand in warm months may be 25% above average and, locally, a doubling of demand can occur on a hot dry day. Higher temperatures would lead to increased demand for water and higher peak demands, requiring increased investment in water resources and infrastructure if restrictions were to be avoided.

- An increase in temperature would increase the demand for irrigation. In times of drought the abstraction for agriculture competes with abstractions for piped water supply by other users.

- A large range of industries use water in production. Some industrial activity (e.g. chemicals production and food processing) would be severely curtailed if water supply were interrupted as a result of drought.

- Increases in the frequency of drought could have an impact on public health through interruptions to domestic and other supplies.

- Climate change may lead to changes in soil structure causing increased leaching, decreased absorption, and changes in agricultural applications of fertiliser and pesticides which could affect river, lake and ground water quality, possibly adversely.

- Many existing power stations rely on river water for cooling. Climate change may reduce the availability of cooling water, particularly during summer months, with possible implications for the operation of individual power stations.

## 7.1 INTRODUCTION AND BACKGROUND

Major changes took place in the water industry in England and Wales in 1989. The water supply and sewage functions of the ten water authorities were privatised and all of the environmental, flood defence and land drainage functions were transferred to the National Rivers Authority (NRA) (WSA, 1990). In addition to the ten water and sewerage companies there are also 29 water only companies which supply about a quarter of drinking water.

The industry is embarking on a greatly increased capital expenditure programme primarily to satisfy EC directives, especially those covering drinking water and bathing water. A new waste water directive from the EC is currently in draft and will require additional expenditure. No allowance was made in the prospectus of the new water companies for any extra expenditure that might be incurred as a result of climate change.

The water service companies are now regulated by a number of bodies, of which the Office of Water Services headed by a Director General and the NRA are the most important in relation to issues raised by climate change. The responsibilities of the Director General include the monitoring of levels of service to customers, i.e. raw water availability, pressure of mains water, interruption of supplies, hosepipe restrictions, and flooding incidents from sewers. Water companies are required to meet agreed targets as part of their licence to trade.

The NRA has a duty to conserve, redistribute, augment and secure proper use of water resources under the Water Act, 1989. The water services companies have a key interest therefore in any resource planning by the NRA.

The industry has to meet European Community directives, of which the drinking water directive specifies the permitted levels of nitrate, phosphates, aluminium, and other substances present in drinking water. Increases in the levels of these constituents in river and groundwater could be very costly for the industry although these costs can be offset by Government action, for example, by control of the sale and use of fertilisers, pesticides and herbicides and the introduction of nitrate and pesticide protection zones.

It can take a decade or more to bring into use a major new water source, thus water demand forecasts are required for up to 25 years ahead. Often reservoir schemes are delayed, scaled down and sometimes rejected as they proceed through the planning system. In the late 1960s demand for treated water was rising at 3% a year; demand was expected to double by the end of the century and many major schemes were investigated by the Water Resources Board (WRB, 1973). There was even research into desalination and a barrage in the Wash estuary by the Central Water Planning Unit (CPWU, 1976a). Subsequently major downward revisions in these forecasts were made as demand rose by only about 1% a year.

The NRA licences abstractions of raw water from inland surface waters and groundwater by industry and agriculture and water supply companies. Demand for raw water by industry has been falling in recent years. The use of non-tidal water by the Central Electricity Generating Board has fallen rapidly with the closure of most inland power stations, so this important sector is now little affected by water shortages. This contrasts with France where some power stations were badly affected by drought in 1990 (BNA, 1990). Abstraction of raw water for agricultural purposes, on the other hand, has been increasing and the spray irrigation element is highly sensitive to dry, hot weather conditions primarily in the lowland south and east of the country. The NRA have identified catchments that are already over abstracted and are conducting research on how to rectify the situation.

Raw water is abstracted by the water companies and treated for public supply. The demand for public water supply overall is rising at 1% per year and is expected to continue rising at about this rate into the next century. Industrial demand for treated water is expected to rise at a much lower rate than that for domestic use. The rises in demand are expected to be lowest in the north and west and highest in the south west, south east and East Anglia, due primarily to demographic change.

In times of water shortage, there are statutory measures (drought orders, hosepipe bans, banning non-essential use of water), voluntary measures (appeals to the public) and demand management measures (temporary mains laying, switching of supplies, etc.) that can delay the imposition of standpipes and rota cuts (IWES/ICE, 1977). The use of these measures is reported to the Director General of Water Services as part of the levels of service reports.

In the last 15 years the industry has experienced three significant droughts (1975/76, 1984 and 1989/90). In each case the drought was followed by very wet spells. Previously, major droughts with water restrictions occurred in 1921, 1934 and 1959. There were also droughts in some groundwater areas in 1942–44, 1962–65 and 1972–74 due to drier than normal winters.

Major drought reports were produced for 1976 and 1984 (CPWU, 1976b; IWES/ICE, 1977; NWC, 1977; WAA, 1985) but not for 1989/90 as yet. Each drought has been substantially different in character affecting water resources differently from region to region. Evidence from these recent droughts may indicate possible conditions in future droughts.

Costing the effect of a drought is difficult, but Gilliland (IWES/ICE, 1977) estimated the cost of the 1976 drought to the water industry as £34.3 million. Due to a combination of demand management to make supplies last as long as possible, plus the eventual arrival of heavy rains, there have been no lengthy and widespread examples of the full economic impact on industry and society. It is assumed, however, that such costs are potentially large.

In 1985, as a consequence of the 1984 drought, a group from the water industry and the Water Research Centre (WRC) met to investigate climate change. The University of East Anglia (UEA) was commissioned to examine rainfall records, both long- and short-term, and to identify any significant trends. A tendency toward greater variability in the weather over the most recent 15 years was postulated based on operational experience. It was also noted that long, hot, dry spells were more likely to end with intense wet spells, as in 1976 and 1984 (1989 eventually followed this pattern), possibly due to higher sea temperatures. The UEA report (Wigley and Jones, 1986) recorded recent increases in dry summers and wet springs but concluded that this was not statistically significant. The report was discussed by the water industry, but did not precipitate action. From 1984 until recently the industry maintained only a watching brief on climate change. The WRC is now sponsoring an expansion of the UEA study and the NRA is allocating resources for research in cooperation with the Institute of Hydrology.

## 7.2 ESTIMATED EFFECTS OF CLIMATE CHANGE AND SEA LEVEL RISE

### 7.2.1 Effects of climate change

*General*

A report from the University of Birmingham (Parry and Read, 1988) considered the wider implications of weather and climate effects on the water industry; the interactions are listed in Table 7.1.

High levels of winter rainfall would at first sight have a generally beneficial effect on long-term water resources in the UK. However, this could be mitigated if there is:

- A longer growing season.
- An increased run-off due to soil breakdown.
- A more concentrated rainfall pattern.
- An increased possibility of a dry winter.
- An impairment of water quality and control as a result of pollution from storm discharges, land run-off and burst pipes. Changes in rainfall affect the deposition of pollutants and pollutant run-off from urban and agricultural land, especially the 'first-flush' pollution following a dry period. These pollution loads could increase markedly if rainfall intensity increases.

High snowfalls have little effect on water resources, but access and operational problems, together with increased risk of power failures and flooding from snow-melt, can result in some additional cost to the industry.

The effects of periods of low rainfall are related to season, duration and location. The costs incurred can be high as water demand increases whilst water quality and source quality decreases. Low river flows result in reduced dilution of domestic and industrial effluents and hence greater treatment costs at water treatment works.

High summer temperatures cause increased water demand and hence depletion of resources, with impairment of river water quality and impacts on pollution control and fisheries. Higher winter temperatures should reduce the number of burst pipes that cause an increase in demand. Operational problems in local water supply and treatment plants should also reduce.

Locally, warmer and drier summers could lead to a strain on smaller 'short-term' resources. This applies particularly to upland sources in the north and west.

Water resource planning has generally been based on the 1 in 50 year return period of river flows and groundwater, using historical records and/or simulated flows using modelling techniques. The duration in months of the critical rainfall period depends on the catchment, the scale of the abstraction and whether the abstraction is from reservoir, river or groundwater. No underlying climate trend is currently assumed. Until the industry is convinced that there is a climatic trend, there is unlikely to be a change in this approach.

*Water Demand*

Demand for treated water in winter, in non-frosty conditions, is fairly constant from day to day. In

**Table 7.1 Impacts and sensitivities of climatic variability on the water industry's activities**

ACTIVITY

| Climate Effect | Water Resources | Water Treatment and Supply | Sewerage Treatment and Disposal | Pollution Control | Land Drainage | Sea Defences | Fisheries Conservation Recreation |
|---|---|---|---|---|---|---|---|
| Rain fall High | L/L | H-M/H-M water quality impaired; farm pollution; storm discharges; pipe bursts | H/H local operational problems | H-M/H-M water quality impaired; farm pollution; storm discharges | H/H flooding | | |
| Drought | H/H related to time of year; duration; rain in previous period and location | H/H water quality impaired; high demand | M/M effluent dilution | H-M/H-M locally important | | | H/L water quality impaired |
| Snow High | L/L 'winter drought' | H/M Site access affected; power failure | H/M local operational problems; site access; power failure | | H/H flooding from snowmelt | | |
| Temperature High | M/M increased demand | L/L water quality impaired | | M/M increased biological activity in rivers | | | H/M distress to fish stocks |
| Low (Frost) | L/L 'winter drought' (snow) | H/M-L local operational problems; power failure | H/L local operational problems | | | H/H surge and wave action | |
| Wind High | H/H security of dams | L/L local operational problems; power failure | L/L operational problems | | | H/H-M locally important | |
| Sea levels High | H-M/H-M Saline intru-; sion; related to location and extent | H-M/H-M water quality impaired | | | | | |

Key: Importance/cost  H=High, M = Medium, L = Low
Source: Parry and Read, 1988

early summer there are significant peak demands (Figure 7.1) during fine, dry weather when the public

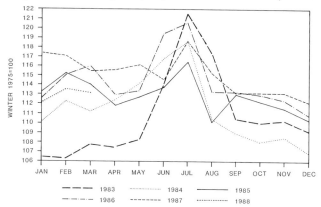

Figure 7.1 Monthly water supply in England and Wales 1983–1988. (Source: Water Services Association).

water lawns and gardens using sprinklers and hosepipes. Water demand during the hot dry summer of 1983 contrasted markedly with that during the cool wet summer of 1987. However, taken over the whole year, a hot, dry summer results in a 2% annual increase in water demand over England and Wales compared with a year with a cool wet summer, although peak demands over a period of a few hours can be great (up to 2.5 times the average) and can result in a pressure drop and occasionally a failure in supply. To satisfy these short duration peak demands fully, a large amount of money would need to be invested in new mains and infrastructure. This is particularly true in the south and east of the country. Local hosepipe bans and further restrictions may also have to be introduced. Domestic water is not generally metered at present so there is no method of restraining demand or recovering the extra cost through tariffs. Currently, metering trials are being conducted throughout the country (WSA, 1989) to examine the cost of metering, the effects of tariffs and their effect on demand. Any conclusions and recommendations are only likely to be introduced in the medium term.

A rise in temperature, combined with sunnier, drier late springs and summers would exacerbate the peak demand difficulties.

## Water Quality

The hot, sunny conditions accompanying the dry weather in 1989 and 1990 has led to water quality problems. Harmful algae developed in several reservoirs, leading to the temporary banning of recreation in some cases. The NRA is now monitoring the situation closely for the first time, and has reported that high temperatures and low river flows resulted in an increase in the total length of polluted sections of rivers in 1989 (NRA, 1990). There was also a major incident of infestation of snails. These phenomena can lead to increased summer operational and treatment costs.

Changes in the climate could induce an unexpected proliferation of undesirable biota as well as increased breakdown of soil structure (see Section 3). Prolonged hot dry periods that are followed by heavy rain may lead to poor surface water quality. This occurred extensively in the peat moorlands of northern England during and after the 1984 drought.

Alterations in forestry and agricultural practice (see Section 5) resulting in changed applications of pesticides, herbicides and fertilisers combined with enhanced leaching of minerals could affect surface and groundwater quality.

## Flooding

High rainfalls, depending on frequency and duration could increase flooding and design criteria may need to be reviewed. At present the projected climate change for the UK, as described in Section 2, does not include details on the level of intensity of short-term rainfall in future.

## Sewerage

Flooding of sewerage systems is strongly related to incidents of short-term intense rainfall and water companies must therefore design sewers to meet the potential return periods of these weather conditions. As yet there is no substantiated evidence of any recent increase in storm intensity or frequency.

## Other effects

Historically, in many UK cities, the water tables were drawn down substantially by local abstractions over many years. When abstractions decline, the water table tends to rise again. Serious problems in urban drainage (Section 11) are already beginning to occur in London, for example, due to the rising water table (CIRIA, 1989). Similar problems are soon to be investigated for other UK cities, including Birmingham and Liverpool. The rising water table has been mainly the consequence of substantial reductions in local water abstraction as a result of the relocation of industry and its more efficient use of water. Wetter winters could result in higher groundwater levels which could have effects also on mining (see Section 9), especially when combined with sea level rise. Climate change might accelerate such changes, with altered river hydrology and sea level rise.

Underground mains and sewers are greatly stressed under conditions of frost and drought when there is soil movement. In general more rapid and extreme changes in either temperature or rainfall will increase the stress on underground assets. In a warmer climate the reduction of bursts in winter may be offset by an increase in summer bursts.

There are implications for reservoir safety if reservoir levels are drawn down more frequently or for longer periods. Embankments could dry out and cause cracking or subsidence problems.

The recent gales of 1987 and 1990 caused short-term and local dislocation of water supply and sewage treatment as a result of power failure. Any increase in the frequency of such gales can be expected to increase such dislocations.

### 7.2.2 Effects of sea level rise

Sea level rise would affect the water industry locally in areas near the coast where saline intrusion in groundwater is already a risk to water quality and where river abstractions for water supply are near the tidal limit (see Section 6).

A gradual rise in groundwater may occur as a result of sea level rise. Saline intrusion in aquifers and wells in coastal areas is a serious problem globally (for example, within the Ganges/Brahmaputra Delta) and may become so in some areas in the UK (primarily in shallow aquifers). A number of coastal towns in the south and east draw water from aquifers close to the sea level limit and some restrictions in use or measures to lessen the effect may be necessary.

The risk of flooding from both high tides and storm surges is likely to increase as sea level rises. In some areas of the UK, notably the East Anglian fenland, where much of the land is at or below mean sea level, pumping is required for water drainage (Beran, 1982). Seepage and salt water intrusion will be greater as the head against which the pumps operate increases. Pumping costs will therefore rise, although the major problem would be maintaining the integrity of levee banks (Beran and Arnell, 1989).

## 7.3 UNCERTAINTIES AND UNKNOWNS

It is not yet known whether recent extreme weather events (droughts in 1975/6, 1984, 1989/90; severe frosts in 1978/79, 1981/2, 1985, 1986; gales in 1987 and 1990) represent a trend to more variable weather. In an industry where managing water has been based on past climate records, any change in climate may mean a mismatch between current design assumptions and the future environment. In some cases this will result in over-provision but in many cases in under-provision.

Total effective rainfall or residual rainfall (i.e. the rainfall left after evaporation) is the most important variable with respect to the water industry. Thus, the industry will require estimates of future seasonal and spatial distribution of both rainfall and evaporation and of any change in the frequency of extremes of residual rainfall over periods of three to 18 months. The industry may also be affected by any changes in local intense rainfall that cause sewage-related and other flooding.

Higher temperatures and associated increased soil moisture deficits could increase ground movements, resulting in a greater incidence of summer burst mains (countered by a reduction of winter bursts). The consequences of ground movement for small water bodies such as reservoirs are even more uncertain. The Building Research Establishment is carrying out research for the Department of the Environment in this area.

Climate change may lead to unknown degrees of change in the use of agricultural chemicals; this in turn will affect the levels of nitrates and pesticides which enter UK soils. The effects of climate change on soils may, in addition, alter the way that these and similar chemicals, such as aluminium, reach water supplies.

The changes in agricultural practice that may be brought about by climate change are likely to alter the requirements for irrigation, especially in the south and east of the country, where the effects of drought could be more extreme. Temperature increases coupled with decreased rainfall in summer may dictate the cultivation of different types of crops with varying water requirements. The effects of these changes on the water industry are not known. However, the hot, dry summers of 1976, 1984 and 1989 may be more commonplace in future, and in each of these years the industry faced varying difficulties.

Any change in energy prices would obviously affect the industry as energy costs, primarily for pumping water and sewerage, are a significant proportion of the costs to the water industry. At £142 million in 1988/89, energy costs represented about 10% of operating costs for the water and sewerage businesses.

## 7.4 PRINCIPAL IMPLICATIONS

Hot and very dry periods of three or four months (1984, 1989) can affect the yield from small upland sources and severely reduce flows in 'flashy' rivers in impermeable catchments. A very dry winter can critically reduce groundwater levels, especially in the south and east of the country. Any increase in the frequency of these kinds of events will be of concern.

A report by Beran and Arnell (1989) has given a hydrological summary of the sensitivities of water resources to climate change. Arnell and Reynard (1989) have developed this summary using historical analogues, regional analyses and hydrological models. They estimated the effect on river flows of a +20% rainfall all year round and of a +20% winter/ −15% summer rainfall in several catchment areas. (This follows the current climate projections given in Section 2). Although initially their study showed that the increased winter rainfall may make up for the summer deficiency, there were several concerns. In the south east of England in particular water resources are highly sensitive to small changes in rainfall and evaporation. Also any increase in frequency of dry winters would have a detrimental effect. Even in upland catchments in the west, hot dry summers can have significant effects on reservoir storage and river flows.

Demand during a warm summer month often increases by 25% above a winter month in the south of the UK, but by less than 5% in the far north of the country. Local areas may experience a doubling of demand on a hot day in a hot dry period (Males and Turton, 1979).

If winters become generally wetter, it is probable that flood risks may increase (catchments being wetter for longer and subject to possibly higher storm rainfalls). The degree of change would depend on catchment type and changes in rainfall characteristics. Many of Britain's most destructive floods have occurred in summer and the extent to which warmer conditions (and surrounding seas) would increase the risk of heavy convective rainfall in summer is unclear.

A large number of industries use water in their production processes. Manufacturing, steel, food, chemicals and oil are among the most prominent. Some industries have installed water storage capacity of about one day on site to ameliorate the effects of interruption to supply; and others have introduced recycling. Nonetheless, a large part of industry would suffer production losses during a period of restricted water supply. The three most recent droughts did not, however, cause major industrial production losses and hence few data on associated costs exist.

The Advisory Council for Agriculture and Horticulture estimated that the demand for water for agricultural purposes would double by the end of the century (ACAH, 1980). This was considered an overestimate at the time by the industry. However, if significant warming occurs, demand for water, for spray irrigation in particular, could rise substantially. At peak demand times the abstraction for agriculture could compete with abstractions for water supply, especially in the south and east.

Possible changes in land use could have an impact on the use of fertilisers (particularly nitrates) and pesticides, with consequent impacts upon surface and groundwater quality.

The considerable health and social implications of an extreme drought must be considered, if algal and other infestations reoccur, if sewerage is disrupted and if standpipes represent the sole water supply in some regions (as was the case in 1976). However, little hard evidence of the actual costs involved exists.

## 7.5 RESEARCH AND POLICY NEEDS

### 7.5.1 Future research effort

Water companies have developed models to estimate water demands and plan water resources. These use historical data on river flows and groundwater levels. In order to test hydrological and water demand, models and sensitivities, the water industry requires accurate projections of rainfall, evaporation and reliability over the annual cycle, and of the variability of extreme effects, before it will move away from using historical information in modelling. Second order sensitivities such as temperature and sunshine should also be considered in any modelling and may be crucial to some issues.

Models which simulate water quality should be developed to take into account the effects of climate change and should be sufficiently sensitive to evaluate a range of impacts over several climate change projections. It is necessary to understand how ecological systems will behave under increased temperature or low flow regimes.

Rainfall and evaporation, and thus run-off, are the primary factors affecting the water industry. Evaporation rates, based on temperature, cloud cover and growing season need to be quantified.

There are many other, individually important, effects caused by changes in other sectors. For example, alterations in soil structure and agricultural practice, especially related to fertiliser and pesticide use, can affect the industry adversely in terms of run-off and groundwater recharge into the water system. Any research in these areas should take note of the effects on water supplies in the UK. In addition, the monitoring of river quality over the long-term at various UK sites should be made a priority.

Those areas and water resources which will be particularly sensitive to the effect of climate change with respect to inundation and salination in particular, should be identified.

There has been no socio-economic appraisal of the potential effect of severe shortages and restrictions of water on industry, agriculture and society. This should be studied.

### 7.5.2 Policy recommendations

The water and sewerage companies are regulated by the NRA which licenses water abstractions and by the Director General of Water Services who monitors adequate levels of service to customers and regulates charges (ultimately the only means of paying for investment). These organisations must allow for the possibility of the occurrence, and the consequent impacts of, climate change upon the water industry. If further resources have to be developed or demand management methods are introduced, it will result in increased water prices. As water resources have been planned using historical records, any change in climate is likely to affect the industry and will cause considerable additional expenditure. In view of the long lead time required to plan, agree and develop water resources, the industry and the NRA require guidance on climate change at the earliest opportunity so that the implications can be assessed.

# Energy

## 8

**SUMMARY**

- The lifetimes of many energy-producing or consuming items of equipment are shorter than the timescales over which climate change might occur. With adequate preparation, successful adaptation to changed conditions is likely.

- Higher temperatures would have a pronounced effect on energy demand. Space heating needs would decrease substantially. Given the present pattern of heating energy supply, natural gas use would decrease most.

- Higher demand for air conditioning would entail greater electricity use. In the longer term the net effect of changed space heating and air conditioning needs could be an increase in electricity demand.

- Changes in building design and technological developments in heating and air conditioning equipment stimulated by climate change could have secondary impacts on energy demand.

- Total UK energy demand is likely to fall as a result of climate change. The effect on energy prices is not known, but is likely to be small. The consequences for different classes of consumer depend on: patterns of use changes in the energy supply system, the rate of depletion of natural resources, and changes in global energy markets.

- The vulnerability of coastal or estuarine energy facilities to storm surges is likely to be increased by sea level rise. All UK petroleum refineries and half of the present power station capacity are located on the coast or on estuaries. There may be a need to strengthen sea defences at certain locations in order to protect against flooding.

## 8.1 INTRODUCTION AND BACKGROUND

The energy sector makes a major contribution to emissions of the greenhouse gases (GHGs) carbon dioxide, methane and nitrous oxide. In particular, more than half of the global emissions of carbon dioxide, the most important greenhouse gas, are attributable to the consumption of energy. Most analysis of the energy sector has focused on its role as a source of greenhouse gases. A significant amount of scenario building work on prospective energy demand and supply has been carried out both at the global (Edmonds and Reilly, 1983) and the national (Dept of Energy, 1989) levels. Equally, the attention of the energy industries is focused primarily on the potential impact on their activities of policies directed at mitigating climate change, including options such as regulatory control or fiscal measures.

In the longer term, climate change would itself have an impact on patterns of energy consumption and on the operations of energy suppliers. However, the literature in this field is comparatively sparse. IPCC Working Group II has identified a limited number of studies, focused mainly on electricity demand and the availability of hydro-electric resources (IPCC Working Group II, 1989).

A full analysis of the relationships between the energy sector and climate would require an assessment, using long-range energy projection techniques, of its role both as a source of GHGs and as a target for climatic impacts. In the absence of any such analysis, a much more limited assessment is made in this section, assuming the climate changes projected in Section 2. Implicit in these climate projections is the assumption that energy demand itself develops on a 'Business-As-Usual' pattern. Should policies be put in place to mitigate climate change, then this implicit assumption would not hold.

An additional difficulty is that the energy sector is likely to undergo major structural changes in the coming decades regardless of policies on climate change. Much of the analysis in this section super-imposes projections of the future climate on the present pattern of energy demand and supply. The limitations of this approach must be recognised. Where possible, the effects of possible changes in the energy sector are indicated.

The energy sector in the UK comprises a wide range of activities which might, potentially, be influenced by a number of climatic variables including: temperature change, precipitation, cloud cover, wind speed, the frequency of extreme weather events, and sea level rise.

## 8.2 ESTIMATED EFFECTS OF CLIMATE CHANGE AND SEA LEVEL RISE

In analysing potential impacts, it is helpful to break the energy sector into four components:

● energy consumption
● the supply of primary energy
● transformation to secondary energy forms, such as electricity or petroleum products
● transport, transmission and storage

This broad categorisation is used as the basis for creating an inventory of potential impacts.

Most of the second component, energy supply, including coal mining and the production of oil and gas, is covered under Section 9 of the report, referring to minerals production. However, a brief description of the possible effects of climate change on renewable energy production is given here.

### 8.2.1 Effects of climate change

*Energy demand*

There is a considerable amount of data available on energy demand and it is possible to approach the question of impacts of climate change from more than one direction. First, it is possible to identify the amount of energy used for specific purposes, such as space heating. Basic physical principles may then be used to estimate the extent to which demands for useful energy might be altered by changes in climatic variables. This may be termed the 'bottom-up' approach.

Alternatively, it is possible to estimate the long-term effect of climate change by seeking analogues with historical responses to climatic variability. 'Top-down' statistical relationships between energy consumption and weather indicators such as temperature, wind, cloud cover and precipitation have been established by supply companies (Parry and Read, 1988).

When it comes to estimating the effects of long-term climate change, there are potential biases in both approaches. A simple 'bottom-up' approach based on measures such as degree day statistics* is likely to overestimate changes in energy demand because it does not account for compensatory responses by

---

*These measure the extent to which mean daily temperatures fall below 15.5°C. Other temperature bases are possible, but 15.5°C, used in statistics published by the Energy Efficiency Office, is judged to be the level at which space heating is required to maintain comfort after allowing for incidental heat gains in buildings.

consumers. For example, higher outside temperatures would effectively lower the cost of providing comfort. Particularly in low-income households, this may lead to some of the advantage being taken in the form of higher indoor temperatures rather than lower fuel consumption.

Equally, the 'top-down' approach may underestimate potential changes in energy consumption. Existing statistical relationships measure the short-term responses of consumers using a given stock of energy-using equipment. The time scale for climate change would allow changes to be made in the specification of, for example, space heating systems. As the result of these changes, a more efficient response to changed climatic conditions would be possible.

Both a simple 'bottom-up' and a 'top-down' approach have been considered in this section. The inherent biases suggest that they may bracket the 'true' response to climate change.

i) 'Bottom-up' approach

Table 8.1 shows an estimated breakdown of final energy consumption in the UK according to end use. This table is based on the quantities of energy delivered to final consumers. Primary energy demand, which takes account of the energy lost in transformation processes such as electricity production, is considerably higher. Those aspects of climate change which might affect each type of end use are identified. Space heating, accounting for 30% of final energy demand, is clearly the most important area. The space heating market is split between households, services and industry in the ratio 61:25:14. Gas dominates space heating with 62% market share (Table 8.2). In the longer term, it is possible that electricity could capture a greater share of the space heating markets.

A good indicator of the underlying demand for useful space heat may be obtained from degree-day statistics. Figure 8.1 shows how monthly degree

**Table 8.1 UK 1988 final energy demand and climatic impacts**

| End Use | Demand (PJ/Year) | % of Total | Likely Change | Climate Variables |
|---|---|---|---|---|
| Space Heating | 1986 | 30 | Less | Temperature, wind speed cloud cover |
| Lighting | 144 | 2 | ? | Cloud cover, precipitation |
| Air conditioning | 24 | 0.4 | more | Temperature, humidity |
| Refrigeration | 97 | 2 | more | Temperature, humidity |
| Transport | 1918 | 31 | — | — |
| Industrial process | 1234 | 20 | — | — |
| Other uses[1] | 963 | 15 | — | — |
| Total | 6276 | 100 | less | — |

Note: 1) water heating, cooking, appliances, motive power
Sources: Energy Efficiency Office sectoral reports

**Table 8.2 Estimated UK final energy demand by fuel in 1988 (PJ/year)**

| End Use | Gas | Electricity | Oil | Coal | Total |
|---|---|---|---|---|---|
| Space Heating | 1170 | 98 | 347 | 281 | 1896 |
| Lighting | — | 144 | — | — | 144 |
| Refrigeration | — | 97 | — | — | 97 |
| Air conditioning | — | 24 | — | — | 24 |
| Transport | — | 12 | 1906 | — | 1918 |
| Industrial process | 463 | 98 | 283 | 390 | 1234 |
| Other uses | 368 | 449 | 90 | 55 | 963 |
| Total | 2002 | 921 | 2626 | 727 | 6276 |

days (averaged over the UK and weighted according to regional energy use) might alter under the estimate of possible temperature change over the period 1990–2050 given in Section 2. Demand for useful space heat could decline by 9% by 2010, 17% by 2030 and 25% by 2050. However, this assumes present levels of demands for comfort and existing building design.

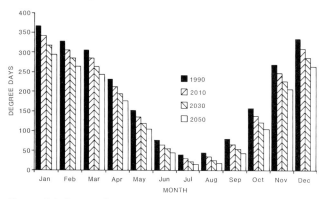

Figure 8.1 Degree days as a result of climate change, 1990, 2010, 2030 and 2050 (for explanation, see text).

The possible changes in construction techniques which might be stimulated by climate change, described in Section 11, could alter consumer responses to temperature variations. Equally, changes in technique could result from policies designed to lower carbon dioxide emissions from energy use in buildings. In addition, the potential for reducing energy demand in existing buildings through simpler retrofit measures is considerable.

The other end uses potentially affected by climate change – lighting, air conditioning and refrigeration – require the use of electricity.

Lighting accounts for 15% of electricity markets. Although demand is likely to be affected by changes in cloud cover or levels of precipitation, there is no certainty about which direction this change might take. The potential for a reduction in lighting demand through the adoption of commercially available technologies, such as miniature fluorescent lamps, is large. Lighting may account for a declining proportion of demand in the future.

Electricity use for refrigeration and air conditioning is even lower. However, energy demand for air conditioning is estimated to have trebled between 1975 and 1985 (Herring, *et al.,* 1988). Approximately a quarter of office area is air-conditioned, though penetration does not exceed 5% in any other class of commercial or public building. Office users now attach a premium to air-conditioned buildings, with the result that air conditioning is now virtually standard in new constructions in urban areas, especially in London and the South East.

The effect of climate change on electricity demand for air conditioning is less easy to predict than changes in demand for space heat. Cooling degree days are a less reliable indicator of energy demand than are heating degree days. It has been estimated by the Building Research Establishment (Milbank, 1989) that a temperature rise of 4.5°C would more than double the average 'full load' usage of a typical air conditioning system, while reducing refrigeration efficiency by about 10%. Thus, air conditioning is a great deal more sensitive to temperature change than is space heating.

If demand for air conditioning were to double from present levels, then it is possible that temperature-driven increases in electricity demand could more than offset reductions in electricity use for space heating. The recent rapid growth of demand for air conditioning has been associated with a high level of construction activity and is likely to drop during the 1990s. The extent to which the growth of demand for air conditioning will resume at a later date is difficult to predict.

While the desirability of air conditioning is firmly fixed in the minds of present office users, there are those in the architecture profession who regard air conditioning as unnecessary and inappropriate in many locations, not least because of the heavy associated demand for electricity and the use of materials, such as CFCs, which have detrimental environmental impacts (Grigg, 1990). On the other hand, climate change may create new markets for air conditioning units in the UK, for example in private dwellings. More research would be needed to gain a better understanding of the possible future role of air conditioning.

ii) 'Top-down' approach

The calibration of most statistical models of energy demand requires that explicit allowance be made for variability in weather. For example, the Department of Energy's forecasting model (Dept. of Energy, 1983) makes explicit allowance for the effect of average winter temperatures (October to April) in the household, commercial and public sectors and for 'other' industry (i.e. excluding iron and steel).

Table 8.3 summarises the effects of the scenarios of future temperature change proposed in Section 2 when combined with the Department of Energy's forecasting equations for each sector.

iii) Comparison of the two approaches

The table also shows results obtained using the 'bottom-up' approach. These were obtained by combining the degree-day changes shown in Figure

**Table 8.3 Possible impacts of climate change – reduction in UK final energy demand by sector[1]**

| | Bottom-up[2] | | | Top Down | | |
|---|---|---|---|---|---|---|
| | 2010 | 2030 | 2050 | 2010 | 2030 | 2050 |
| Household | 3–9% | 7–16% | 11–22% | 4–6% | 5–9% | 7–11% |
| Service sector | 3–8% | 7–15% | 11–21% | 5–8% | 7–12% | 9–15% |
| Iron & Steel | <1% | <1% | <1% | — | — | — |
| Other industry | 1–3% | 2–4% | 3–6% | 1–2% | 2–3% | 2–4% |
| Agriculture | — | — | — | — | — | — |
| Transport | — | — | — | — | — | — |
| Total | 1–4% | 3–7% | 5–10% | 2–3% | 3–5% | 4–6% |

Note: 1) Based on the pattern of final energy demand in 1988
2) Assuming changes in space heating energy demand only

**Table 8.4 Possible impacts of climate change – reduction in UK final energy demand by fuel due to changes in space heating needs[1,2]**

| | 2010 | 2030 | 2050 |
|---|---|---|---|
| Gas | 3–8% | 6–14% | 10–20% |
| Coal[3] | 2–5% | 4–9% | 7–13% |
| Coal[4] | 1–3% | 2–5% | 4–7% |
| Oil | 1–2% | 2–3% | 2–5% |
| Electricity | <2% | 1–3% | 2–4% |
| Total | 1–4% | 3–7% | 5–10% |

Notes: 1) Based on the pattern of final energy demand in 1988
2) Based on the 'bottom-up' approach
3) Effect on coal demand in final energy markets only
4) Effect on coal demand allowing for changes in power station markets – assumes all changes in electricity demand are reflected in coal use

8.1 with the space heating requirements for each sector. There is a good correlation between the patterns of demand reduction suggested by the two techniques.

However, the 'bottom-up' approach leads to higher estimates of the reduction potential for energy demand, particularly in the household sector. This may be due to the fact that people allow the internal temperatures of their homes to rise in warmer weather. This factor is reflected in the 'top-down' estimates but not in the 'bottom-up' ones.

iv) The effects on fuel markets

Table 8.4 shows how potential reductions in final energy demand might be distributed around the different fuels. Gas would be most significantly affected, reflecting its dominance for space heating.

The electricity estimates in Table 8.4 reflect only reductions in space heating demand. As noted above, increases in power use for air conditioning and refrigeration might more than compensate for space heating in the longer term.

Although coal's markets in the final demand sectors could be significantly altered by climate change, the main market for power generation would be less affected. Table 8.4 shows the potential effect on total coal markets assuming that the coal industry bears the full effect of changes in electricity demand.

Table 8.4 can only be indicative as it is based on patterns of energy demand in 1988 and major changes in energy markets are inevitable over the coming decades.

v) Patterns of Demand and Energy Prices

As well as changing total energy demand, climate change is likely to alter temporal patterns of demand. For example, the changes in the degree-day pattern shown in Figure 8.1 would, assuming no change in temperature variability, reduce total space heating energy demand more than it would peak demand, resulting in a lower utilisation of space heating systems. Increased use of air conditioning would, on the other hand, reduce the large difference between the winter and summer demands for electricity resulting in a better utilisation of power stations.

Changing demand profiles have implications for fuel choice. For example, the lower utilisation of space heating systems could favour electricity, the use of which entails lower capital costs than does gas. Equally, gas displaced from space heating markets might become more attractive in other markets such as power generation.

Energy tariffs would change in order to reflect altered patterns of demand. However, the nature of these changes is unclear, as much depends on the way in which the energy sector evolves. Price changes attributable to climate change are likely to be dwarfed by those caused by general market conditions or, possibly, by fiscal policies aimed at reducing GHG emissions.

### Energy supply

The effects of climate change on fossil fuel supply are described in Section 9.

Climate change could have an important impact on the availability of cost-effective power from renewable sources. The main renewable potential in the UK is believed to lie with wind, tidal energy and biomass.

The potential for the use of hydro power would be affected by changes in precipitation patterns. However, hydro electricity makes little contribution to UK electricity generation and the possibility of further large schemes being developed is minimal, though there is some potential for small run-of-river projects.

The UK enjoys good wind resources, though there are considerably differing views about the size of the potential (Chester, 1988; Grubb, 1989). It is not yet possible to project the effect of climate change on wind speeds, but these will have a significant effect on the degree to which wind resources are economically viable. It has been estimated that there are 7200 km$^2$ of land in the UK with wind speeds above 8 m/s, at which wind power is approaching economic viability, and a further 4500 km$^2$ with wind speeds above 7.7 m/s (Chester, 1988).

There are two major prospects for tidal energy generation in the UK, on the Severn (7,000 MW) and Mersey (700 MW) estuaries. The development of these projects is uncertain given the long lead times and high capital costs. The barrages would have a design life of 120 years and sea-level rise and possible changes in weather conditions would need to be taken into account in developing such projects.

There are also implications for the potential productivity of biomass, particularly that based on short rotation forestry, though this is not currently practised in the UK.

The potential for solar energy in the UK would be affected by changes in cloudiness. It is not yet possible to make predictions about these. Geothermal energy would be unaffected by climate change.

### Energy transformation

The generation of electricity might be affected by climate change. Many of the present coal-fired power stations in England rely on river water for cooling (Table 8.5). However these tend to be smaller, older stations many of which are unlikely to

**Table 8.5 Location of UK steam power stations in operation and under construction: number of sites (GW capacity)**

|  | Coal[1] | Oil | Nuclear | Total |
|---|---|---|---|---|
| Sea/Lower Estuary | 13 (14.0) | 9 (13.3) | 10 (11.4) | 32 38.7 |
| Upper Estuary | 6 (11.3) | — — | — — | 6 (11.3) |
| Lake | — — | — — | 1 (0.4) | 1 (0.4) |
| River | 18 (15.2) | 1 (0.1) | — — | 19 (15.3) |
| Canal | 1 (0.2) | 1 (1.0) | — — | 2 (1.2) |
| Total | 38 (40.8) | 11 (14.4) | 11 (11.8) | 60 (67.0) |

Note: 1) Includes dual fired stations

Source: Annual Reports and Accounts – CEGB, SSEB, NSHEB, NIE

remain in operation for long. To the extent that climate change might affect precipitation and hence the availability of cooling water, it is conceivable that the operation of individual power stations might be affected. The output of some French nuclear reactors was restricted during 1989 due to lack of cooling water. Cooling water availability may be diminished during the summer months.

Electricity generation would also be affected to a very minor extent by the effect of ambient temperature changes on turbine efficiency.

*Transport and transmission of energy*

Coal and oil are transported by road, rail and sea. The implications of climate change on transport are discussed in Section 12.

The transport of oil and gas by pipeline and electricity by cable is a special feature of the energy sector.

Electricity transmission in the UK is primarily by means of overhead lines. Transmission capability varies from one season to another as a result of temperature changes. The thermal rating of a typical 400 kV double-circuit line falls from 2720 MVA in winter to 2190 MVA in summer (Eunson, 1988). Climate change would therefore make a small, though appreciable difference to transmission capability. In terms of costs, this could be reflected as: a requirement to reinforce transmission links, higher operating costs through running power stations out of merit, or changing the siting of proposed power stations. Actual costs would be highly dependent on the precise development of the electricity supply system.

Overhead power cables are vulnerable both to violent weather and to icing in winter. The storms of October, 1987 resulted in an average loss of 250 minutes supply to customers in England and Wales (Electricity Council, 1987/88). Apart from the additional costs of repairing lines, the value of the power lost to consumers may be considerable. The electricity supply industry uses £2/kWh as the value of lost load (VOLL) in establishing reserve capacity requirements, implying a cost to consumers in October 1987 of approximately £200m. Such storms may become more frequent as a result of climate change.

### 8.2.2 Effects of sea level rise

A significant number of major UK energy facilities are located at coastal sites. The vulnerability of any particular site to storm surges and the need, if any,

to strengthen barriers to rising waters requires assessment on an individual site basis. A coastal location does not necessarily imply that a site is untenable.

All 14 UK oil refineries are situated on coastal sites. At the same time 32 power station sites, comprising more than half of the UK's generating capacity, are situated either on the sea or on lower estuaries (Table 8.5).

In considering the potential effects of sea level rise, careful consideration would need to be given to the timescales for change. Power stations have lifetimes of 30–40 years and, in principle, future investment could be directed away from coastal sites. New power station designs, such as combined-cycle gas turbine plant, are less dependent on cooling water and hence coastal or river siting will become less critical than it has been in the past.

The energy industry is mature and there is a great deal of infrastructural investment, such as transmission lines, road and rail access, associated with existing sites. Economic incentives to utilise this investment, coupled with potential local resistance to establishing new greenfield sites, means that there may be considerable inertia with respect to power plant location.

Given that major onshore oil production in the UK is unlikely, petroleum refineries are likely to remain coastally sited.

### 8.3 UNCERTAINTIES AND UNKNOWNS

Comparatively little work has been carried out on the effects of climate change on the energy sector. However, analytical tools are available and many of the uncertainties, particularly those on the demand side, could be narrowed considerably by further research. An accurate assessment of many of the energy sector impacts is dependent on a better understanding of changes in climatic variables other than temperature and sea level rise, notably cloudiness, precipitation, wind speeds and frequency of extreme weather conditions.

### 8.4 PRINCIPAL IMPLICATIONS

Most of the impacts of climate change on the energy sector are of an incremental nature. It is not meaningful to identify thresholds for changes in climatic variables beyond which unmanageable challenges are posed.

The lifetimes of many energy-producing or consuming items of equipment are considerably shorter than the timescales over which climate change might occur. The expected lifetime of boilers or air conditioning units, for example, might be of the order of 10–20 years. With adequate preparation, successful adaptation to these changed conditions should be possible.

An exception to the rule about the incremental nature of change could be the effect of sea level rise on particular coastal or estuarine energy facilities.

Though change is likely to be manageable, the impacts on energy demand in particular may be large. For natural gas, climate change could depress demand increases (or reinforce demand decreases) by up to 0.3% per year. This remains, however, well within the range of energy demand changes which have been managed in the past.

There are important interconnections between energy and other sectors. As described in Section 11, changes to building design could have important impacts on levels of space heating and air conditioning energy demand. Meanwhile, the price and availability of energy have important implications for a range of sectors including manufacturing, water and mining.

It is not possible at this stage to say how energy prices might change in response to climate change. Broadly speaking, UK energy costs are likely to decrease as a result of reduced space heating needs in the early stages of the next century. However, in the longer term, increased use of air conditioning may push costs up. Price rises in response to altered demand profiles are possible, particularly for natural gas. Changes in electricity prices are less easy to predict.

Although energy prices as a whole might rise, different classes of consumer may be affected in different ways depending on their load profiles and on tariff structures.

## 8.5 RESEARCH AND POLICY NEEDS

### 8.5.1 Future research effort

The following research priorities are suggested:

i)  Efforts should be made to establish the direction and magnitude of possible changes in secondary climatic variables such as cloud cover, wind speed, precipitation and frequency of extreme weather events.

ii) The impacts of climate change on energy demand need to be assessed within a scenario framework which allows the consideration of a wide range of possible developments over the next 50 years. Such an analysis is a prerequisite to any serious consideration of the energy price impacts of climate change.

iii) The impacts of climate change on the availability and cost of renewable energy sources need to be assessed.

iv) The vulnerability of individual coastal and estuarine energy facilities to sea level rise deserves investigation.

v)  The possible effects of climate change on demand for air conditioning need to be assessed.

### 8.5.2 Policy recommendations

Energy suppliers have an important role to play in identifying in more detail the possible impacts of climate change on their activities. Both Government and energy suppliers should consider taking explicit account of climatic effects in carrying out long-range projections of energy supply and demand. Climate change may have a small but perceptible effect on the evolution of energy demand, even over the next 20–25 years. To the extent that climate change poses strategic challenges, Government could play an important role by stimulating research and informal discussion of the longer-term issues. Changes in building design and the role of renewable energy resources are areas in which Government may have a particular role to play.

# Minerals Extraction

## 9

**SUMMARY**

- Projected changes in temperature and rainfall are unlikely to result in significant technical problems for the oil and gas, coal, industrial minerals and aggregates industries in the UK. Minerals are extracted successfully throughout the world under a wide range of climatic conditions and appropriate working practices and technology could be adapted to the UK. In addition, UK engineers and management would have adequate time to develop any new techniques that may be required to accommodate the likely impacts.

- Rises in sea and estuary levels and changes in the water table and salinity would, in some areas, be very important to land-based mineral extraction and dredging. There could be increases in pumping and mineral processing costs, and the restoration of extraction sites would be more complicated.

- The projected rates of rise in sea and estuary levels are sufficiently low to allow investments in current operations to be recovered provided that there is no increase in the intensity and frequency of storm surges.

- Decisions to improve sea and river defences would require increased UK production, and possibly the increased import, of larger sizes of hard stone. For example, recent damage to sea defences at Towyn, in North Wales, gives an indication of the impact of storm surges. Boulders weighing up to five tonnes were required for repair and these are larger than usually quarried.

- Changes in the energy market in connection with space heating and air conditioning requirements due to temperature changes would affect the demand for coal, oil and natural gas and would imply changes in the structure of the UK minerals extraction industry.

## 9.1 INTRODUCTION AND BACKGROUND

The UK domestic resources of coal, oil and natural gas, hard rock, sand and gravel and other industrial and metalliferous minerals are an economically significant sector of the UK economy.

The total value of UK minerals production in 1988 was £15.3 billion and direct employment was about 160,000. Crude petroleum exports made a major contribution to exports; the trade balance in selected minerals by value in 1988 is shown in Table 9.1. Since the 1985 value of UK minerals production of £27.5 billion, there has been a decline due to lower oil prices, changes in the exchange rate and loss of oil and gas output from fields connected to the Piper Alpha platform. Also, the cost of coal to the electricity generating industry has fallen substantially in real terms. In contrast, the value of construction materials in 1988 and industrial minerals produced in the UK rose by approximately 9% from 1987 to reach the level of £1.8 billion (British Geological Survey, 1989).

For all practical purposes UK coal reserves are owned by the British Coal Corporation. They have a duty to work coal except for their ability to grant limited licences. Coal is extracted by both underground and surface mining techniques and, although the industry is dominated by British Coal, there is a small private sector operating under the licences. Coal is also recovered from old mine dumps and slurry ponds. Total annual production is currently about 100 million tonnes with just over 80% of the output obtained by underground mining. British Coal, which accounts for around 96% of UK coal production, had 73 underground mines and 57 surface mines in March 1990. The private sector had a large number of very small underground mines and open pit operations at the end of 1989. The principal market for coal in the UK is electricity generation, which consumes 80% of production. Industry uses 9% of the output and is the second market in order of importance and the remainder is supplied for coke ovens and domestic heating (British Coal Corporation, 1989).

Significant deposits of lignite, estimated at 1,000 million tonnes have been discovered in Northern Ireland. The location of the coal and lignite deposits are shown in Figure 9.1.

Offshore crude oil production in 1988 was 109 million tonnes and land production was 0.9 million tonnes. Natural gas liquids, including condensates, recorded a 1988 output of 5 million tonnes and offshore natural gas production was 38 million

tonnes oil equivalent. Estimates by the British Geological Survey (BGS) for 1989 are 87 million tonnes of crude oil, 4.4 million tonnes of natural gas liquid and 37 million tonnes of oil equivalent for natural gas. Additional output of natural gas results from coal mining together with some other land sources (British Geological Survey, 1989). The location of oil and gas production together with oil refineries and gas terminals are shown in figure 9.1.

The known technically recoverable reserves of coal in the UK can be measured in centuries at present levels of production (Whittaker, 1990). The proved plus probable reserves for oil and natural gas are 13 years and 32 years respectively (Taylor, 1990).

In 1987, there were approximately 3,000 quarries and pits associated principally with the production of hard rock, sand and gravel and other mineral production such as china clay, slate, gypsum and metalliferous ores (BACMI, 1989). The British Geological Survey estimates for 1989 output are: limestone (excluding dolomite) – 112 million tonnes;

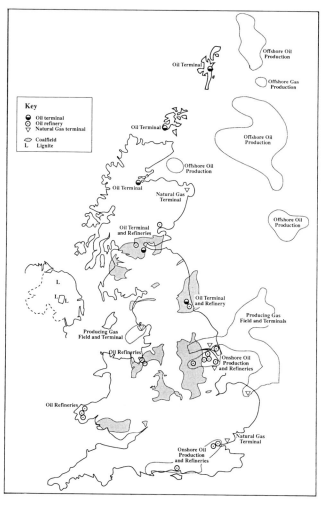

Figure 9.1 Distribution of coal, oil and natural gas in the UK. (Source: British Coal Corporation, 1989 and British Geological Survey, 1989).

**Table 9.1  Trade balance in selected minerals by value, 1988**

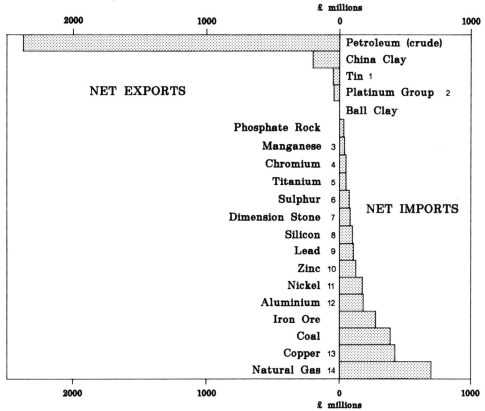

1 Concentrates and unwrought metal
2 Concentrates and unwrought platinum etc.
3 Ore, ferro-manganese and unwrought metal
4 Ore, ferrochrome, metal and oxides (est.)
5 Ores and concentrates, slag, ferro-titanium and metal
6 Crude and unroasted pyrites
7 Building and monumental stone and slate

8 Elemental silicon and ferro - silicon
9 Concentrates, bullion and refined metal
10 Concentrates and unwrought metal
11 Matte, ferro-nickel, refined (unalloyed) nickel
12 Metallurgical alumina and unwrought metal
13 Unrefined copper and refined (unalloyed) copper
14 Liquified and other

**Output and Value of Production, 1988**
**Total Value £15 293 million**

| Mineral | Coal | | Natural Gas | Natural Gas Liquids | Crude Petroleum | Iron Ore | Tin | Common Sand and Gravel |
|---|---|---|---|---|---|---|---|---|
| | Deep | Opencast | | | | | | |
| Production | 83 461 | 17 924 | 66 768* | 5 004 | 109 402 | 224 | 3.4 | 136 404 |
| Value | 4 268 | | 1 956 | 282 | 6 933 | 1 | 14 | 543 |

| Mineral | Limestone and Dolomite | Igneous Rock | Sandstone | Chalk | Common Clay and Shale | China Clay | Ball Clay | Fuller's Earth |
|---|---|---|---|---|---|---|---|---|
| Production | 125 680 | 51 960 | 18 901 | 14 516 | 18 899 | 3 277 | 716 | 213 |
| Value | 498 | 200 | 68 | 38 | 26 | 224 | 21 | 14 |

| Mineral | Salt | Silica Sands | Potash | Fluorspar | Gypsum and Anhydrite | Miscellaneous Minerals | Silver | Gold |
|---|---|---|---|---|---|---|---|---|
| Production | 6 130 | 4 340 | 767 | 104 | 4 000** | 1 950.3 | 67 944*** | 167*** |
| Value | 39 | 46 | 58 | 10 | 24 | 25 | 5 | |

\* Coal Equivalent   \*\* Based on 1989 estimate   \*\*\* troy oz         Source: British Geological Survey (1989)
Production in thousand tonnes
Value in £ million

73

dolomite (excluding limestone) – 21 million tonnes; chalk – 15 million tonnes; sandstone – 20 million tonnes; igneous rock – 55 million tonnes; land and marine sand and gravel – 135 million tonnes. Marine sand and gravel in 1988 was 20 million tonnes. Employment levels in the aggregates industry are about 33,000 (BACMI, 1989).

In recent years there has been a small increase in certain areas of metalliferous mining in the UK. The Parys Mountain orebody in Anglesey contains copper, lead, zinc, silver and gold and the development of a new underground mine is well advanced. Small gold deposits have been discovered in Northern Ireland and Scotland and the deposits in Scotland are being developed. The underground tin mining industry in Cornwall has recently been reduced as a result of very low prices for the product.

Additional sources of UK mineral production are china clay, gypsum, rock salt, fluorspar, barytes and potash. The production of gypsum, china clay and potash has been increasing. Fluorspar has decreased, due partly to the reduction in CFC manufacture and partly to changes in steel making. Rock salt production has been reduced owing to lower demand in the recent milder winters.

Figure 9.2 shows the distribution of aggregates and the location of some of the deposits of metalliferous ores and industrial minerals.

## 9.2 ESTIMATED EFFECTS OF CLIMATE CHANGE AND SEA LEVEL RISE

### 9.2.1 Effects of climate change

Mining operations and oil and gas extraction are carried out in a wide range of climatic conditions throughout the world. The extraction industry has shown itself to be capable of providing the technology and human effort and skill to overcome considerable physical difficulties.

It is reasonable to assume that any variations in temperature, winds, dry periods, wet periods, overcast skies and barometric pressure which are envisaged by the projected climate change can be accommodated, as these would take place over a long period of time and give adequate oppportunity for making the necessary adjustments. Also, the time scale envisaged provides a sufficient period, in most cases, to recover the investment in the operations. Long life underground coal mines have production terms of over half a century, but there are also numerous surface mines and quarries which have a lifespan of only a few years.

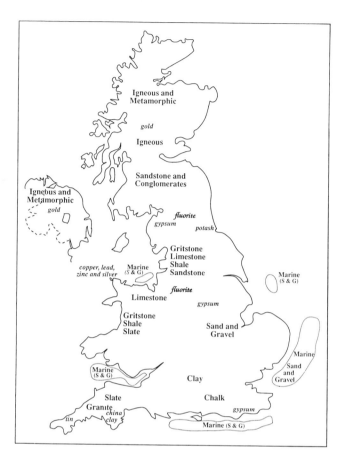

Figure 9.2 Distribution of aggregates and mineral deposits in the UK (Source: British Geological Survey, 1989, and BACMI, 1989).

Adjustments to climate change could be made by incorporating techniques which have been successful in similar circumstances in other parts of the world. It is also safe to assume that the UK and international mining machinery manufacturers, together with innovative management, could develop new technologies and techniques if these were necessary.

The problems which could occur as a result of climate change can be classified more accurately as inconveniences and irritations which, although causing rises in costs, would not be sufficient to significantly alter the financial and competitive position of the industry. It is likely that climate change of the type projected could lead to improved conditions in some mining operations in the UK.

The mining industry in the UK, particularly the surface mining sector, has a large amount of experience in dealing with the problems associated with noise, dust, traffic congestion and restoration. Many of the surface coal mining operations are near built-up areas, thus environmentally acceptable techniques must be demonstrated before planning permission is granted.

Long, hot summers are likely to increase environmental problems due to increased volumes of dust. Additional dust suppression will be required, particularly for surface mining. Stockpiles of coal and waste at deep mining operations would have to be kept to minimum levels. Dry periods culminating in a water shortage would detrimentally affect coal washing facilities, with consequent production losses and increases in costs. Overcast skies can affect the strength of blast noise due to reflection from the cloud base.

Many UK underground mines are working seams which are located within strata that contain quantities of methane under pressure. Under normal operating conditions, gases are gradually released from these strata into the ventilation airstream, which is designed to be sufficient to dilute them satisfactorily. Normally the pressure difference is relatively stable but when barometric pressure falls rapidly, this can result in a dramatic increase in the gas emission rate which would temporarily interrupt production. Current trends in underground coal mines are towards working fewer faces at very high rates of production with high availability time for machine operation. Interruptions in continued production on these faces can result in increasing costs which affect the financial returns from the capital intensive equipment. This problem can be alleviated by methane drainage from the strata direct into pipe systems. This technique is widely practised in UK mines and the methane is utilised on colliery premises or elsewhere for power generation and space heating.

Surface mines for all minerals would be adversely affected by increased wet periods; flooding of the pits and pollution control difficulties would be the main consequences. Large underground mines are currently working below the water table and the possibility of flooding is minimal. The effects of increased pressure on shaft linings would need to be studied. Some small underground coal mines of the type which are currently owned by the private sector would be more detrimentally affected.

For the quarrying of construction materials, the most important impact would be a movement of the water table. A raised water table would increase pumping costs at a hard rock quarry. In the case of sand and gravel, excavation under dry conditions is favoured but water is essential for processing. A higher water table could complicate restoration on quarry sites as soils used for restoration can only be moved in dry periods. The stripping of soils prior to extraction could be severely impeded by unseasonable rain.

Oil and gas from the Northern Basin of the North Sea is produced in some of the most challenging climatic conditions found anywhere in the world. Oil and gas exploration and production technologies have improved rapidly as a result of movement into more difficult operating regions e.g. the move from the Southern Basin to the Northern Basin of the North Sea. It is unlikely that technology would fail to keep pace with the envisaged climate change scenarios.

The minerals extraction industry will have to respond to any changes in construction standards which are developed to deal with the effects of climate change.

The above analysis has been confined to the direct impacts of climate change on the UK minerals extraction industry. It is important to be aware that indirect effects of climate change, such as the possible imposition of carbon taxes and fuel switching in order to cut carbon dioxide emissions, would have a significant effect on the industry. Coal production could be reduced and natural gas production could increase. Changes of climate may require increased water storage and distribution facilities, resulting in an increased demand for construction materials.

### 9.2.2 Effects of sea level rise

As with the impact of climate change, the indirect effects of sea level rise are likely to have a much greater impact on the UK minerals extraction industry than the direct effects.

Increased sea level would not affect the physical operation of marine dredging of sand and gravel because the depth of extraction has gradually increased over the years. Increased processing costs are likely to be incurred if chloride levels rise and, under extreme conditions, increased sedimentation could lead to extra contamination by clays. Additional coastal erosion could also result in remobilisation of the sand and gravel deposits.

Estimates are required about the extent of the areas of the UK likely to be affected by sea level rise both in the coastal and low lying inland regions, before any analysis of the direct impact on mining and quarrying can be attempted. The time scale of the changes and the probability of policy decisions to defend the coastline are also relevant.

If extra sea defence work is required there may well be a need for more UK hard rock quarries to be equipped to meet the special demand. It may also be necessary to increase the size of quarries and planning permission problems could result in the locality of the operations. Sea defence requires large stones and coastal quarries are usually used to meet such demands. Depending on the amount of sea defence work to be carried out, it could prove necessary to promote the production of large stones from UK quarries to avoid the import of large sizes (e.g. from Norway). It might be practicable to use materials from colliery spoil heaps for these construction works.

Recent damage to sea defences at Towyn, in North Wales gives an indication of the major effort required to carry out the necessary rebuilding. Boulders weighing up to 5 tonnes were needed and these are larger than usually quarried. In addition it is also necessary to provide significant quantities of ready-mixed concrete for the reconstruction task (Anon., 1990).

## 9.3 UNCERTAINTIES AND UNKNOWNS

In order to quantify the impact of climate change on the UK minerals extraction industry, information is needed on the following: changes in the water table, the extent of coastal erosion, sea defence policies, modification to construction standards, changes in energy demand, frequency of barometric pressure changes and lengths of periods of dry and wet weather.

## 9.4 PRINCIPAL IMPLICATIONS

The overall direct impact of changes in temperature and rainfall in the UK could be both negative and positive for the minerals extraction industry. The negative impacts can be summarised as irritations and inconveniences. Current and developing technology and management flexibility should be able to accommodate the likely range of adjustments resulting from these climate changes. In addition, many of the problems will already have been solved by mining operations in countries with climates similar to the type envisaged for the UK. Individual operations could be affected to a greater or lesser extent but no major problems are anticipated.

Rises in sea level and estuary levels and changes in the water table and salinity are important to land-based mineral extraction and also to marine dredging. The extent of the overall changes, the frequency of surges in sea and estuary levels and the timing of the changes are critical to assessing the impacts. The timing and extent of the projected rises in sea level would provide adequate time to recover investment in existing operations.

The impact of policy decisions to climate change and sea level rises are of paramount importance to the minerals extraction industry. The UK coal industry will be sensitive to attempts to reduce carbon dioxide emissions by a reduction in the burning of coal for electricity generation. Changes in the energy market in connection with space heating and air conditioning will have an impact on the UK coal industry (Section 8). A decision to improve sea and river defences will require increased production or imports of large sizes of hard rock.

Public opinion and general knowledge levels about the domestic and international energy markets will influence Government decisions.

## 9.5 RESEARCH AND POLICY NEEDS

### 9.5.1 Future research effort

Future research effort should be directed towards the impacts of sea level rise on coastal and estuary sites where there are facilities and operations for extracting, processing, refining and storing of minerals. Movements of the water table should also be investigated to determine their effects on the minerals industry. A study of the availability of materials and the costs for improving sea defences is also important. Construction standards may need to be altered to accommmodate climate change and research should also be directed to energy conservation and new coal burning techniques. Further analysis of all the greenhouse gases to provide an improved emissions data bank is a research requirement; the levels of nitrous oxide emissions from agriculture relative to those from fossil fuel combustion should be quantified.

### 9.5.2 Policy recommendations

Policies on the extent of sea and river defences are essential in order to assess the impact of climate change on the minerals extraction industry. Emission standards should be defined as well as the question of emission permit trading, both internally and between countries. The role of natural gas and nuclear power for future electricity generation in the UK must be clarified and the question of carbon taxes addressed.

# Manufacturing

# 10

**SUMMARY**

- Manufacturing industry contributes approximately 30% of the UK's Gross Domestic Product and accounts for over 35% of the UK's income from exports. It is therefore an important sector to consider in relation to the impacts of climate change.

- Most of manufacturing should be able to adapt to climate change. Climate change will produce conditions in the UK which already exist elsewhere in Europe where manufacturing industry contributes effectively to GDP and competes internationally.

- The major concern for manufacturing industry, 40% of which is located in coastal and estuarine areas, relates to the potential for sea level rise. For rates of rise up to about 5 cm per decade the costs could probably be absorbed, but rates in excess of this and over the longer term could substantially increase costs. These impacts will differ greatly from site to site.

- Increases in temperature, with reduced winter freeze, would be likely to improve productivity in some areas such as the building industry, to improve the transport of raw materials and finished products, and to decrease disruptions in production due to weather. However, the food industry could be affected by locational changes in produce supply and may find an increased need for refrigeration.

- Adverse effects may arise in certain manufacturing industries due to shortage of water (e.g. paper making, brewing, food industries and power generation). Repeated annual droughts could adversely affect their production.

## 10.1 INTRODUCTION AND BACKGROUND

In 1987 the manufacturing industry in the UK contributed 30% to the UK's Gross Domestic Product (GDP) and the export of goods accounted for over 35% of the UK's income from the exports of goods and services (Central Statistical Office, 1989). It is therefore a major focal point for the potential impacts of climate change.

Other sections of this report cover the services which manufacturing companies need, e.g. water supply, energy, raw materials and transport. Industry requires a reliable and economic supply of these services. In addition, the location of a company is influenced by a number of factors, e.g. location of the market for the product, the availability of a competent labour force, good transport accessibility and the proximity to raw materials.

All the factors mentioned above will affect the company's product and its price and therefore have a major impact on a company's performance and competitiveness both nationally and internationally.

In defining the parameters affecting manufacturing business it is clear that any climate change resulting in changes in temperature, precipitation or sea level could have some direct, and indirect, influences on British manufacturing. Most of these will be difficult if not impossible to quantify in the short term. However, they will need to be taken into account in future strategic planning. It is important to set this against the projections for the UK climate for c. 2050 which may result in a temperature rise of 1–3°C. This would produce a new climatic environment, but is similar to that already existing in parts of Europe and in which industry operates and is able to compete successfully on international markets. It is therefore likely that most changes will be ones which industry can adapt to and indeed such changes may well bring benefits to some industrial sectors, provided that they have anticipated the possible changes in sufficient time.

## 10.2 ESTIMATED EFFECTS OF CLIMATE CHANGE AND SEA LEVEL RISE

Tables 10.1 and 10.2 summarise the contribution of various industrial sectors to UK GDP and address some of the impacts on sectors sensitive to climate change and the possible responses.

Table 10.1 Contribution of sectors of business to UK GDP

| Industry | % GDP |
|---|---|
| Manufacturing (including Energy and Water Supply) | 30.09 |
| Construction | 6.38 |
| Agriculture, Forestry, Fishing | 1.63 |
| Transport and Communication | 7.24 |
| Distribution, Hotels and Catering | 13.67 |
| Financial Services | 18.44 |
| Others | 22.55 |

Source: Central Statistical Office, 1989

### 10.2.1 Effects of climate change

Other sections in this report consider the impacts that climate change could have on many of the factors which are important in providing a base for manufacturing industry. These are impacts on the following:

*Energy Supply:* Industry and commerce require continuity of supply at a minimum price. The costs of supply interruptions can be high even though, on average, energy represents only about 3% of business costs. In energy intensive industries (e.g. steel, chemicals, cement, paper), this can rise to 13% (Table 10.3). A breakdown of energy use in the UK industrial sectors, including iron and steel, is given in Table 10.4. The chemicals sector accounts for 20% of industrial energy use (6% of the UK total energy consumption of 60 billion therms). Climate change will affect the provision of energy in terms of consumption, transformation to secondary forms, transmission and storage (Section 8). Increased costs of energy supply as a result of emission tariffs imposed are also likely. Another point to note is that the climate change described in Section 2 is a result of the emission of greenhouse gases, 50% of which is $CO_2$; about 27% of UK $CO_2$ emission sources are industrial (Department of the Environment, 1989).

*Water Supply and Sewerage:* The overall projected increase in winter rainfall is relatively small ($3\pm3\%$ by 2010, $5\pm5\%$ by 2030 and $8\pm8\%$ by 2050); however, warmer, drier summers could lead to a strain on short-term water resources locally (Section 7), with concomitant impacts on those industries which use large amounts of water, e.g. paper making, the food industry and brewing. Figures show that in 1989 industry used more than 30% of the water supplied by the water companies (Water Services Association, 1989). Abstractions under such conditions may also be restricted. Industry was responsible for 32% of the total water abstracted in

**Table 10.2 Impacts of climate change on manufacturing industry**

| INDUSTRY | % of GDP (1987) figures | Potential Impacts | Potential Responses |
|---|---|---|---|
| Food, Drink and Tobacco | 3.03 | Temperature rise/increased rainfall resulting in increased rate of food deterioration and change in crop production patterns. Temperature rise may be beneficial if crop production is increased | Requirement for increased refrigeration. Increased crop production may reduce raw material costs. Increased temperature – higher demand for drinks. |
| | | Reduced rainfall resulting in reduced freshwater availability and concentration of pollutants | Lower crop production may increase raw material costs and require the import of more raw materials. |
| | | Sea level rise leading to threats of inundation of coastal sites and reduced land for raw material production or factory sites. | Increased water use efficiency and tighter pollution controls. |
| | | | Sea level rise may require the movement of factories to higher ground |
| Chemicals, and Man-made fibres | 2.54 | Decreased rainfall/sea level rise resulting in reduced freshwater availability. Need for tighter controls on effluents if lower river flows. | Greater minimisation/recycling of process water. Greater investment needed to control aqueous emissions at source |
| | | Increase in sea level could result in flooding of coastal sites. | Investment in sites protection and drainage. Movement of sites inland. |
| Paper Printing and Publishing | 2.24 | Increase in sea level/reduced rainfall could result in water shortage and flooding of coastal sites. Changing climate could impact on timber and thus wood production. | Sources of wood pulp may change. Recycling/minimisation of water use. |
| Engineering | 5.81 | No specific impact. | – |
| Textiles, Clothing and Footwear and Leather | 1.58 | Increase or decrease in temperature will alter public demand for clothes, etc. | Industry will adapt and mirror markets demands. |
| Aerospace | 0.88 | No specific impact | – |
| Vehicle Manufacture | 1.32 | No specific impact | – |
| Rubber and Plastics | 1.01 | No specific impact | – |
| Others | 11.68 | No specific impact | – |

**Table 10.3 Total UK energy purchases as a percentage of estimated production costs (1984 figures): some examples from industry.**

| Industry | Estimated Production Costs (£m) | Total Energy % |
|---|---|---|
| Bricks & Ceramics | 1409 | 13.19 |
| Cement, Lime & Plaster | 2655 | 11.47 |
| Glass & Glassware | 1140 | 13.19 |
| Basic Chemicals | 5013 | 9.72 |
| Paper & Board | 1830 | 10.18 |
| Iron & Steel | 6849 | 7.86 |
| Water Supply | 1155 | 11.40 |

Source: Department of Energy, 1989b

**Table 10.4 Industrial energy use, 1988**

| Industry | Million therms |
|---|---|
| Chemicals | 3339 |
| Iron and Steel | 3235 |
| Mineral Products | 1627 |
| Food, Drink and Tobacco | 1597 |
| Mechanical Engineering | 1047 |
| Paper, Printing & Publishing | 890 |
| Vehicles | 700 |
| Non-ferrous Metals | 604 |
| Textiles, Leather & Clothing | 519 |
| Electrical Engineering | 484 |
| Construction | 475 |
| Unclassified | 769 |
| Other | 1557 |
| Total | 16842 |

Source: Department of Energy, 1989a

1985. In addition, depletion of water resources during periods of increased temperatures will result in less dilution of industrial wastewaters and hence a greater need for treatment at source, to reach the levels required by Water Authorities prior to discharge into the public sewerage system.

It is worth noting that studies of the impact of the 1975/76 drought on industrial productivity revealed a pronounced decline in some sectors (notably ferrous metals and the utilities which use large quantities of water) and increases in others (particularly clothing, footwear, drink and tobacco which all experienced higher retail sales) (Palutikof, 1983).

*Raw Materials:* Warmer winters will improve productivity in the building industry (Section 11) and the transport of raw materials and finished products may be eased (Section 12). Food processing and related industries such as brewing could find themselves faced with changes in the types and rates of supply of raw materials, as will those industries dependent upon forestry. Section 5 outlines the impacts of climate change on agriculture and forestry, which will necessarily impinge on the food and wood utilising industries. An increased need for refrigeration of perishables to combat spoilage may also result from higher temperatures.

*Transport Routes:* The movement of both raw materials and finished products will be subject to the impacts of climate change on transport networks (Section 12). Any changes in transport costs are likely to be reflected in the costs of manufactured products, which will in turn affect markets. Any reduction in the severity of winters would result in a positive impact on transport efficiency.

*Productivity:* Improved winters could lead to improved productivity though this may be partially offset in other parts of the year by reductions in productivity in periods of extreme high temperature. The overall effect is likely to be small but positive.

It is also worth noting that certain industrial sectors showed a marked decline in productivity during the harsh 1962/63 winter, notably bricks and cement, and timber and furniture (Palutikof, 1983).

*Markets:* There will be slowly changing patterns of demand for products such as clothing and drinks due to the shift in average temperature. The education of the public towards the possible effects of climate change may, however, induce a more marked alteration in lifestyle and product preferences, as has already been seen in the influx into the market of CFC-free and recycled products. Manufacturers and suppliers will have to adapt to meet these changes and any costs incurred. However, it is unlikely that climate change would lead to rapid changes in market demand, certainly not in comparison with normal market variations. Manufacturing industry would therefore be able to adapt to such changes.

*Communications:* If the frequency of storms such as those experienced in October 1987, and in the winter of 1989/90 were to increase, disruption of communications would become more frequent, with consequent impacts on markets and supply lines.

### 10.2.2 Effects of sea level rise

Approximately 40% of UK manufacturing industry is located on or around the coast and estuaries (for reasons of transport, access to cooling water, etc.). Most of this industry is therefore naturally concerned about the possibility of sea level rise and the increased frequency of extreme conditions. However, since the projections for sea level rise are only 20 cm by 2030 and 30 cm by 2050, it is likely that the majority of establishments would be able to adapt.

The key issue is that of site location. The appropriate balance of protection and relocation will need to be found; this will depend on the sea level rise expected. Relocation of industry would be extremely expensive and would have its own major impacts on the environment, especially since it would include the workforce as well as the installations. There is a need to carry out detailed local assessments of the potential impacts. Although the scale of these impacts is not likely to be great, they need to be quantified.

Increases in the frequency and severity of storm surges may cause structural damage to coastal constructions and necessitate protective measures. These will also disrupt or damage transport routes, power transmission lines and water supply.

Salination of groundwater may also result from a rise in sea level; in those coastally-located industries where abstraction is practised, this could cause difficulties.

### 10.3 UNCERTAINTIES AND UNKNOWNS

The potential impacts on the manufacturing sector have considerable degrees of uncertainty attached to them. If industry is to take action to respond to changes in climate it is essential that both strategic and immediate impacts are researched in depth.

In particular the timescale of these possible impacts must be evaluated in order to determine whether industry could adapt effectively. Potential changes in markets and manufacturing costs will alter the competitiveness of many businesses, particularly at the international level. There is a need to examine how the major competitors of the UK will also be affected. It is possible that many of the impacts will assist British companies; however, the issues will need to be studied carefully so as to limit the negative impacts and maximise the positive ones.

In summary, the net effect of climate change on manufacturing industry would be the result of many direct and indirect impacts. Because many factors are involved, it is extremely difficult to estimate the outcome.

## 10.4 PRINCIPAL IMPLICATIONS

The above sections have summarised the way in which manufacturing industry is critically linked with most other sectors which will be affected by climate change. However, it will be the rate of change which will be most critical. Businesses, particularly heavy manufacturing, usually have a long return on investment. It would therefore be changes in return due to changes of climate that would have the greatest detrimental impact on manufacturing. In particular the direct effect of an increase in sea level on the viability of current coastal industrial sites and their surrounding hinterland will perhaps be the most critical factor.

## 10.5 RESEARCH AND POLICY NEEDS

### 10.5.1 Future research effort

Very few data exist to enable climatic impacts on the manufacturing industry to be quantified. It is imperative that greater resources are devoted to studying both the greenhouse effect and its potential impacts on, e.g. population migrations, lifestyle alterations (i.e. changes in consumption patterns, modes of travel) all of which will affect industry.

Attempts must also be made to quantify the costs of measures designed to limit the greenhouse effect and the costs of measures designed to cope with the changes which it may bring about (Confederation of British Industry, 1989a, 1989b).

Priority areas for research in relation to impacts of climate change on manufacturing include:

- The impact of sea level rise on the availability of industrial sites.
- The impact of sea level rise on water supplies and water disposal.
- The impact of climate change on raw materials, in particular agricultural produce.

Particular issues to be addressed in these research areas include:

- Identification of target industries at risk from climate change.
- Quantification of the climate change parameters of key importance.
- Identification of thresholds of these parameters above which significant impacts (positive and negative) could arise.
- Evaluation of the rates at which various industrial sectors could adapt to the possible ranges of climate change parameters currently being proposed.

The potential effects of climate change on the manufacturing industries of other countries should be examined to allow the assessment of impacts on competition.

### 10.5.2 Policy recommendations

Emphasis should be placed on quantifying alternative methods of economically accelerating measures to ameliorate or prevent climate changes by, for example, encouraging energy efficiency or by introducing penal measures. Ultimately, any action taken will probably be a combination of preventative and remedial measures.

# Construction

## 11

**SUMMARY**

- Warmer winters would reduce heating energy demands in buildings by significant amounts (about 30–40%).

- Warmer winters would confer some benefits for construction productivity, as the industry would become less affected by snow and ice.

- Increased winter rainfall would adversely affect building operations.

- Unless buildings are better designed to take account of warmer summers, interiors could become significantly warmer in summer. This is likely to increase the demand for air conditioning and the consequent use of electrical energy.

- Buildings are vulnerable to extreme climatic events such as high winds and severe wind-driven rain, but it is currently not possible to assess changes in long term risks due to climatic extremes other than temperature extremes. However, some risks are likely to decrease, for example, snow loading.

- The greatest negative impacts on construction are likely to arise from the combination of sea level rise interacting with alterations in inland water hydrology. The impacts will be localised in vulnerable areas.

- Foundation stability on shrinkable soils would be affected by increased winter rainfall combined with drier summer soil conditions. South eastern England will remain the area with most properties at risk.

- Design risk codes, currently based on historical climatic experience, will in future need to match the assessed risks emerging from the impacts of climate change.

## 11.1 INTRODUCTION AND BACKGROUND

The field of construction, which is very weather sensitive, embraces architecture, building and civil engineering. It also includes the building materials supply industries, which provide a significant proportion of construction added value, but whose role is often overlooked. The availability of cheap, reliable, weather-resistant building materials able to stand up to the imposed climatic forces is important. In addition, constructions need to be durable and safe, and economic to build and operate. They must also be operationally maintainable at reasonable cost, otherwise they rapidly become more vulnerable to weather. Large scale civil engineering projects have long lead times, sometimes as much as 15 to 20 years.

Outdoor climate acts on indoor climate. Indoor climates in constructions must be kept suitable for the activities to be housed in them. This weather-sensitive process requires substantial inputs of energy for heating and lighting, and also sometimes for cooling. If changes of climate occurred, these would be likely to alter the balance between heating and cooling demands.

The construction industry is involved in work both above ground and below ground on a wide range of terrains, some sheltered, others very exposed. A variety of special constructions of various kinds are needed at the land-sea interfaces. The construction industry also plays an important role in the development of river and other hydrological works, dams, water supply systems, and provides works for safeguarding against hydrological risks, such as river floods and urban drainage failures. Other works are required to secure water quality. There are important cross links with water policy and coastal defence policy. The industry also provides the 'track support' above and below ground for land-based transportation systems (roads, railways, gas and electricity power lines, etc.). The industry constructs the ports, the built interchanges between water-borne and land-borne transportation systems, and also between air-borne and land-borne transportation systems. All these activities are weather sensitive.

## 11.2 ESTIMATED EFFECTS OF CLIMATE CHANGE AND SEA LEVEL RISE

### 11.2.1 Effects of climate change

*Impacts on design*

The problem of assessing the impacts of likely climate change on construction falls into two distinct parts:

(i) *The assessment of the likely impacts of climate change on existing constructions:* This requires the identification of any consequent changes or modifications needed to counter any unacceptably adverse effects of climate change. Changes may be needed to counter effects for a number of reasons, for example, to make existing constructions safer or more comfortable for people indoors or more economic to operate, including being more energy efficient on a year round basis, and/or less productive of carbon dioxide, oxides of nitrogen and CFCs. The correct time to make the alterations also has to be decided. One would reach different timing conclusions for responding to climate change if the predicted changes occurred linearly or exponentially or as a sudden jump with time. If changes of climate occurred, these would be likely to alter the balance of heating costs and cooling costs in existing buildings. Some preliminary estimates of the impacts are given in Table 11.1.

(ii) *The assessment of how current design practices might require modification:* This assessment should also include an assessment of the design changes desirable to reduce the future production of greenhouse gases and other adverse pollutants. The economics of modifying different elements of existing constructions to meet increased predicted climatic loads will depend very much on the positions of the elements to be altered and their interconnections with the rest of the structure. Altering foundations below ground can be extremely expensive. Refixing stronger cladding elements is far less expensive. It may not prove economic to modify some structures to cope with climate change. Fortunately, providing the workmanship is sound and adequately durable materials have been used, there is a considerable safety reserve in the design of the majority of existing UK structures, for example, against total wind failure. However the effective safe economic life of existing constructions is very dependent on proper maintenance. Extreme climatic events often bring the final write off of poorly designed, constructed or maintained structures.

In considering the design of new constructions, one must note that the design life of the basic structure is normally relatively long. Account therefore may need to be taken of estimated climate 100 years hence, if unacceptably expensive mid-life alterations are to be avoided. The service systems however, such as heating, cooling and lighting, have shorter design lives, typically of 20 to 30 years. In this subsector, thinking about the design impacts of climate change to 2020 may be entirely adequate, provided an awarenesss of possible future change leads on to keeping appropriate future reservicing options open. External finishes, such as paints, may only have a design life of around 5 years, and the choice of suitable protection can be treated as a present weather decision. However, finishes manufacturers generally deal with quite long lead times for new products.

*Climate change, risk analysis and design codes*

Construction design and the contracting process climatically are very much dominated by issues of safe design procedures. Climatic risk analysis forms an important part of design. Key risks include high winds, snow loading, driving rain, thermal expansion, and excessive rates of weathering. In the UK, wind is especially important (BRE, 1989; Cook, 1985).

In spite of serious gales in recent years, current observations do not support statistically any trend towards greater windiness in the UK (Hammond, 1990). The long term scenario concerning extremes of wind remains ambiguous, and it is currently not possible to make any objective statements. Nor is it possible on present information to predict changes in driving rain patterns. It is likely, if the climate became more dominated by Atlantic type weather, that driving rain especially on the west side of the UK would increase.

The most important impacts of climate change on design codes are likely to be those linked with changes in extreme values of wind, rainfall and snow, for example, the wind loading code (BSI, 1990) and the snow loading code (BSI, 1988). Temperature loads applicable to some structures, claddings, bridges, railway lines, etc., will require review.

*Impacts on site work*

The construction process is very sensitive to weather. The industry tenders for work taking careful account of regional variations in weather. Rain, snow, high winds and frost all impact adversely. Increased rainfall in winter would impact on productivity, both through its direct influence on working, and also through its indirect influence on the state of the ground. A warmer winter climate would reduce the impacts of frost on the construction process. More high winds would lead to the need for greater attention to be paid to wind protection during construction. Contractors would use their experience to modify their tendering and contracting methods to match the changes experienced in regional climates. The industry is capable of evolving its practices to adapt to gradual climate change. However, as contractors have to work to defined Codes and Standards, which are developed by rather slow consultative processes, it will be important for the industry that the issue is addressed in advance of climate change. The design of machines and equipment used in construction may require modification. For example scaffolding may require modification to resist greater wind loads. Existing work on the variability of the impacts of present climate on the construction process might provide a basis for examining future climate impacts on site production (BRE, 1990).

*Weathering and climate change*

The processes of weathering of building materials are very complex, and it is proving difficult enough to develop satisfactory weathering prediction models in existing UK climates, far less developing models to assess the effects of future climate change on weathering (Building Effects Review Group, 1989). The key weathering factors vary with material. Seven main categories of construction materials may be identified:

- Porous inorganic materials based on silicates, e.g. bricks, ceramic tiles, concrete.
- Porous limestones, other building stones
- Bulk organic materials, e.g. transparent plastic sheets, asphalts, bitumens.
- Metals.
- Protective coatings, often pigmented organic films, e.g. paints.
- Inorganic glazing materials, e.g. glass.
- Timber.

With the first two porous groups of materials, cycles of freeze and thaw are important, as is the presence of salts in the pores. An increase in temperature, as part of any change in climate in the UK, would be likely to reduce frost damage.

*Acid rain and the weathering of building materials*

While acid rain is a key factor in the weathering of many building materials, it is difficult on current knowledge to predict the impacts of future changes in the outdoor climate and of changes in acid rain concentrations on the durability of different building materials (UK Review Group on Acid Rain, 1987). The prediction of damage becomes especially difficult when there is more than one acid gas present. For example the synergistic damage effects of $NO_x$ and $SO_2$ are not currently understood. However increases in the winter rainfall climate could lead to greater amounts of acid driving rain impinging on buildings, which would accelerate damage. Higher relative humidities in association with more driving rain could lead to more frequent saturation with acid rain. Increases in relative humidity could influence the time corrosive unevaporated acidic rainfall remains in cracks and crevices. Greater damage to buildings would result.

If there was more westerly type weather in winter, more chlorides would be deposited, especially in inland areas. However, offsetting this factor, less salting of roads might be necessary, thus reducing risks of spray-induced corrosion on roadside structures, and on construction adjacent to highways.

A drier summer climate would be broadly a less corrosive climate, but a brief period of heavy rain after a prolonged dry spell in which there has been dry deposition of particulate pollutants can produce a highly acidic solution, which causes severe damage to buildings. As rain may sometimes beneficially dilute concentrations of adverse pollutants on building surfaces, the damage caused by infrequent heavy rain in a dry summer may actually exceed that caused by a higher but steadier rainfall. In general, the chemical reactions occurring in corrosive processes proceed more rapidly as the temperature rises. The present estimates of future climate changes are too vague to assess changes in risks.

Changes in carbon dioxide increases the 'natural' acidity of rainfall. The increased gaseous concentration could directly affect the carbonation of concrete.

It is important that present research aimed at elucidating the interactions between outdoor air quality, rainfall, other climatic factors and building materials damage be sustained, so that, when more precise long range climatic scenarios can be prepared, the capacity for estimating the consequent damage will exist.

*Within-ground construction based impacts of climate change*

Stable and durable foundations are essential for all forms of construction. Soil movements due to wetting and drying of expansive clay soils can cause extensive foundation damage which is expensive to rectify (Boden and Driscoll, 1987). There are large areas of the UK where summer drying out of clay soils is responsible for serious building damage, especially widespread if a dry summer follows a dry winter. Tree roots are often responsible for the extraction of large amounts of water from around foundations, therefore potential changes in the movements of water into and out of the ground are extremely important for construction.

While the climate changes currently projected for the UK include wetter winters, it is not yet known whether summers will become wetter or drier. If climate change results in wetter winters and drier summers, the scale of damage due to this cause will increase, especially in south east England. As the rate of growth of trees could increase, given the carbon dioxide fertilization factor, the root structures could become more extensive and damaging. Increased clay soil movements could also cause much damage to drainage systems and to other buried services.

Warmer winters would reduce frost heave risks, but, with increased rainfall, the winter state of ground could become more adverse for construction processes. Frost would penetrate less far into the ground, opening the eventual possibility of reducing the depth of cover necessary for adequate frost protection of services and foundations.

The position of the water table is very important, especially in areas where extensive use is made of tanked basements. Rising water tables caused by the decline in abstractions are already causing serious problems (CIRIA, 1989) (Section 7). There are also issues concerned with the bearing capacity of foundations in waterlogged grounds with certain soil types.

Increased rainfall rates would have an impact on building and urban surface drainage problems. Standard procedures for urban drainage design might need to be reviewed.

The consequent land-use planning implications of climate change for construction seem to be that certain lower lying areas of the UK could become more vulnerable to flooding risks, and to substantially raised foundation risks, including saline-induced corrosion of foundations and buried services.

*Infestive biological ecosystems and construction*

As buildings, especially houses, contain much timber and timber-derived products as part of their construction, they can be vulnerable to insect and fungal infestation. Building joinery, for example, is especially vulnerable to fungal decay. This vulnerability would increase were the amount of driving rain impinging to increase. Note must be taken of the implications of the need to move away from non-sustainably grown tropical hardwoods to conserve the tropical rain forests. This change has important implications for future choices in joinery. Cold winters discourage the development of certain timber-destructive insects. Warmer summer climates could favour species like the house long-horn beetle, a timber destructive species which is currently quite common in the slightly warmer climates of countries to the south of the UK.

Moisture conditions are very critical in determining the risks of fungal attack in timber products. If there was more rain in winter, and relative humidities also rose, the risks of structural fungal attack, in certain situations, could increase. The safety against attack of timber joists embedded in externally exposed construction, which is just on the favourable side of the balance margin at present in the UK, could tilt to the other side of the margin, causing relatively frequent failures due to fungal attack. The need for better timber protection using acceptable preservatives and paints is likely to increase, to counter a potentially more aggressive biological environment.

## Impacts on heating, cooling and lighting

The most obvious beneficial impact of global warming in the field of construction will be a reduced demand for winter space heating because the number of winter heating degree days should decrease (Table 11.1.) Millbank (1989) estimates, for an average centrally heated house, temperature rises of 1.5°C to 4.5°C would reduce heating requirements by between 15% and 45%. His estimates for non-domestic buildings are similar. However, if the winter climate becomes windier, then there will be some offsetting losses due to increased infiltration. One also has to consider the impacts of changes in the average short wave radiation, which might decrease, and in long wave radiation, the net exchange of which might be considerably altered due to changes in cloudiness. Solar energy is a climatic energy supply source used for heating and daylighting. Any fall would offset the fuel saving gains due to higher mean air temperatures. The net outgoing radiation may be lower in winter due to increased cloudiness. All these changes place special importance on the role of improved daylighting design, improved artificial lighting

Table 11.1 The estimated effects of a 4.5°C increase in mean temperatures on annual energy costs in existing buildings and annual carbon dioxide production from buildings in the UK (Milbank, 1989), with interpolated preliminary scenario estimates for the smaller temperature rises projected for 2010, 2030 and 2050, (estimated by J.K. Page).

| MILBANK DATA | Value (£m) per year | Million tonnes of $CO_2$ |
|---|---|---|
| Housing: | | |
|   Heating Fuels | 1,200 saving | 27 saving |
| Non-domestic: | | |
|   Heating Fuels | 1,200 saving | 25 saving |
|   Refrigeration | 150 increase | 3.5 increase |
|   **Balance:** | 2,250 saving | 48.5 saving |
| Percentage change | 15% saving | 15% saving |
| Saving per degree rise | 3.3% saving | 3.3% saving |
| **SCENARIO ESTIMATES** | | |
| **2010** | | |
| N. Scotland: +1.1°C | | |
| S. England: +0.8°C | | |
| Assumed weighted mean: +0.85°C | 2.8% saving 425£m | 2.8% saving |
| **2030** | | |
| N. Scotland: +2.6°C | | |
| S. England: +1.6°C | | |
| Assumed weighted mean: +1.8°C | 5.9% saving 883£m | 5.9% saving |
| **2050** | | |
| N. Scotland: +3.9°C | | |
| S. England: +2.3°C | | |
| Assumed weighted mean: +2.6°C | 8.5% saving 1,283£m | 8.5% saving |

Note: Milbank's values are indicative, being based on preliminary studies. A Business-As-Usual scenario for insulation standards was assumed. A 1°C change in average temperature in the UK is roughly equivalent to a 10% reduction in energy consumption in centrally heated domestic dwellings, which currently form about 70% of the UK domestic stock. For non-domestic use, the relationships are more complicated. Page has interpolated from Milbank's values weighting S. England as 5 and N. Scotland as 1. In view of the present small impact of refrigeration in the totals, no special adjustments were made for this.

controls and more intelligent window glazing systems (Fancheotti, 1990; BRE, 1988). The building heating energy impacts of climatic warming are discussed in more detail in the energy section of the report (Section 8).

Most UK buildings are naturally ventilated. Mean daily summer temperatures indoors in such buildings are invariably significantly above those outdoors. A summer mean daily outdoor rise of 2°C would make a greater proportion of naturally ventilated buildings unacceptably hot. Substantial investment would be needed on solar screening systems to protect windows. The need for more effective control of internal heat gains from lights and appliances would increase. The summer impacts will be greatest in lightweight buildings due to the high peak indoor temperatures. In contrast, in buildings of high heat capacity, the diurnal swing of indoor temperature is considerably damped. The indoor diurnal swing is about one third of the outdoor temperature swing. This damping helps reduce summer thermal discomfort during daytime. Higher vapour pressures in conjunction with higher outdoor air temperatures would have particularly adverse impacts on indoor comfort and health (Page, 1990).

Even with a mean temperature rise as small as 2°C, the number of cooling degree days will rise very sharply in the UK, because current UK summer outdoor temperatures are often at or below the acceptable thermal comfort level. Milbank (1989) estimates that the average full load air conditioning use would more than double to 2500 hours per annum with a mean temperature of 4.5°C. The energy consumption of existing air conditioned buildings would increase, and plant sizes might have to be increased to cope. The energy implications of changes in air conditioning are discussed in Section 8 of this report.

### 11.2.2 Effects of sea level rise

The construction industry will be affected by the need for development of new sea defence works and the improvement of existing ones. Sea level rise and changes in storm patterns would also make demands on the reconstruction of port facilities.

Saline water intrusion could produce severe corrosion problems in existing underground elements of construction. Adverse effects could occur in the drainage systems near to coastal areas.

As sea level rise occurs, the underground water problems below cities located close to estuaries could become more acute, especially if saline intrusion is also encountered.

## 11.3  UNCERTAINTIES AND UNKNOWNS

The key uncertainties for the construction industry lie in the limited usefulness of current estimates expressed in terms of changes of climatic means. The construction industry attaches great importance to engineering risk analysis. There is too much at stake to do otherwise. The problem of forecasting extreme values in climate change needs to be addressed with far greater scientific effort. Present relatively unresearched efforts for guessing extremes could lead to risk assessments based on very dangerous assumptions about future climatic statistical distributions. The construction industry is likely to continue a conservative approach on risk.

## 11.4  PRINCIPAL IMPLICATIONS

The construction industry would find itself in an extremely difficult situation if a major stepwise adverse climate change were to occur. The construction response time for altering existing constructions is relatively long. There are work load limitations to coping with sudden change and also dealing with the mitigation of the construction effects of widespread disasters. If change is steady, accepting the underlying variability of historic climate to which the construction industry is well accustomed, the construction industry would respond to well-defined goals. The construction industry is likely to continue to see British Codes of Practice and British Standards as key documents affecting their commercial security. Design decisions, made at specific historic times, may be undermined by climate change, threatening future public safety and security; this is a critical issue. There will also be indirect impacts from climate change because the construction industry is highly interactive with almost all sectors of the economy. In the near term the below-ground impacts could present the greatest difficulties, especially in estuarine regions. Land use planning could become a very sensitive issue.

## 11.5  RESEARCH AND POLICY NEEDS

### 11.5.1.  Future research effort

The most important research need is to develop better means of using GCMs for risk analysis purposes. There is a need to define the desirable outputs required for risk analysis in the construction industry, and to communicate these needs to climatic modellers in very clear terms.

More research on UV radiation is clearly desirable to clarify the various impacts on building materials, as well as the underlying climatological trends. The scientific standards need to be raised to achieve more reliable UV measurements in the UK; it is very desirable to establish an accurate specto-radiometric record of changes taking place in the UV spectrum. Such a record might head off much current ill-informed speculation concerning the predicted adverse impacts of UV radiation due to atmospheric ozone decreases.

Further studies should be carried out on probable changes of estimated heating, cooling and daylighting costs consequent on climate change in the UK.

In view of the importance of summer overheating issues, special steps should be taken to strengthen research and development concerning hot weather cooling problems in buildings, giving particular attention to natural cooling techniques rather than placing excessive reliance on the expanded use of air conditioning systems. Expanding research and development work in this field would both create immediate benefits and lay the preparatory foundation to cope with the greater range of summer overheating problems likely to be encountered in construction in the UK by the year 2050.

There is scope for better research coordination of work relating to the summer overheating problem between different Government research agencies, trade associations and academic groups concerned with building energy conservation. The role of site microclimate design in particular deserves greater emphasis as an economic means of substantially increasing summer performance at very low cost compared with air conditioning.

The role of new technologies for building needs to be examined in the context of climate change. For example, advances in the technology of glazing (e.g. electrochromic glasses) potentially could prove very effective in the moderation of summer conditions. A systematic study of their likely all year round impacts is needed to ensure manufacturing efforts are directed towards the most effective specifications for promoting good year round performance, both visual and thermal.

Interactions between insurance industry experience of climate damage, the assessed degree of terrain exposure and construction improvement potential should be systematically investigated.

## 11.5.2 Policy recommendations

An appropriate scale of research and development is needed on the impacts of climate change on the construction industry. The Building Research Establishment has made a promising initial start. This effort needs to be fostered and expanded. Trade-based research associations should be encouraged to identify with this expanded effort, concentrating especially on those impacts most relevant to their specific industrial interests.

Measures should also be taken to encourage collaboration between the various Research Councils, especially NERC, ESRC and SERC in the development of policies for research to interlink studies of the construction-based causes of climate change with the consequent climate change, its specific impacts on the construction industry and the technological, social and economic changes needed to adapt appropriately to predicted change.

NERC, ESRC and SERC should, in discussion with the insurance industry, consider a research policy on climatic aspects of construction risk management. The insurance sector should be encouraged to interact more vigorously with the governmental and academic research institutions on climatic damage and control research.

The Planning Directorate of the Department of the Environment should start to review the land-use planning issues at stake in climate change with the aim of providing improved guidance to local planning authorities on the economic control of climate change risks affecting construction. Future construction settlement patterns will need to be reviewed, and appropriate affordable policies for either safeguarding the future of existing constructions or relocating them, will need to be evolved.

In types of construction where the consequences of failure for public safety are high, for example sea defence walls, periodical engineering reassessment procedures might need to become a mandatory process controlled by regulation.

The British Standards Institution, in conjunction with the Meteorological Office and the Building Research Establishment urgently need to review the procedures by which Code development and Code revision for the construction industry can take better account of informed UK and European scientific predictions concerning potential climate changes which affect Code preparation and revision.

In order to reduce risk to foundations on shrinkable clays, policy regarding landscaping close to buildings in vulnerable areas may need to be reviewed to ensure that, in such areas, a greater priority is given to keeping trees away from structures. Reducing risks to structures from trees blown down during gales requires more attention in landscaping.

As many of the below-ground problems likely to emerge from climate change are rather specialised, involving interactions between river hydrology, urban hydrology, oceanography, civil engineering and soil mechanics, there would be an advantage in setting up a special subgroup to report on the impacts of climate change on construction issues arising from within-ground climate change.

A greater policy effort is needed to persuade the construction industry that new buildings could be better designed to optimize indoor comfort for all the senses (visual, thermal, air quality and auditory) with low energy consumption on an all year round basis, rather than a winter-alone energy conservation basis. This would help safeguard the future summer performance of buildings under a warmer climate. Further steps should be taken to make building designers more aware of the merits of passive solar heating, natural daylighting and natural cooling, both to reduce greenhouse impacts and to offer improved indoor environments at acceptable costs. In conservation policy, the role of daylighting as an energy savings technique should be combined with proper respect for traditional and new task-related visual comfort requirements.

**SUMMARY**

- Sensitivity to weather and climate change is high for all forms of transport, but especially for road and air transport.

- A reduction in the frequency, severity and duration of winter freeze would reduce disruption to transport systems. Winter maintenance expenditure on UK roads should decrease, with less salting and snow-clearing required. A decrease in the number of freeze-thaw cycles would result in reduced road damage.

- Snow and ice present the most difficult weather-related problems for the railway system: these problems could be expected to be reduced, with potential savings in locomotive rolling stock design, point heaters and de-icing equipment. Increases in temperature would also reduce snow and ice problems that hinder aircraft operations, although changes in prevailing winds could affect runway operations or re-distribute noise impacts in urban areas.

- Any increase in the frequency of severe gale episodes could increase disruption from fallen trees, masonry and overturned vehicles, and interrupt flight schedules and airport operations.

- Sea level rise and more frequent coastal flooding could cause structural damage to roads, bridges, embankments and other transport infrastructure.

- If precipitation increased, this could exacerbate road flooding, landslips and corrosion of steelwork on bridges.

- Changes in the demand for some goods, e.g. perishable foodstuffs, coupled with higher ambient temperatures, may affect the pattern and frequency of distribution of goods to wholesalers and retailers.

- Changes in demand for travel, such as increased leisure journeys, may result from perceived 'better' summer weather.

## 12.1 INTRODUCTION AND BACKGROUND

Efficient, rapid, dependable transport is a prerequisite of an advanced economy and interruptions, disruptions and dislocations to any mode of transport will quickly have 'knock-on' effects across a very wide range of industrial and commercial activity. A clear division can be recognised in the nature of likely impacts of climate change into:

(i) Changes of short term extremes of climate, many of which can cause dangerous travelling conditions and lengthened journey times or can impose additional costs on transport operations.

(ii) Changes of the frequency of conditions which, although not extreme, can be hazardous. A good example might be changes in frost and icy road frequencies.

The high degree of sensitivity of UK tranport to a wide range of weather variables has been explored by Perry (1981), Parker, *et al.* (1986) and Parry and Read (1988). From contemporary climatic impact assessment studies, models of the likely scale of future impacts based on changes of climate projected will be considered.

Ironically, road transport is a serious contributor to global warming. Over the last 5 years traffic has increased by 27% and by 2025 an increase of between 83% and 142% is expected, resulting in increases in carbon dioxide emissions by between 62 and 113 million tonnes per year.

## 12.2 ESTIMATED EFFECTS OF CLIMATE CHANGE AND SEA LEVEL RISE

### 12.2.1 Effects of climate change

Tables 12.1–12.3 summarise the recent deliberations of a small panel of climate experts convened by Parry and Read (1988). Taking the information from Table 12.1 and 12.2 together it can be seen that a rank order of impacts can be recognised ranging from least impact (sea transport) to greatest impact (air transport). On the land surface, rail transport is

**Table 12.1 Ranked first-order magnitude of impact on UK Transport: operating procedures**

| Variable | Low cloud | Snow | Rain | Ice | Fog | Wind |
|---|---|---|---|---|---|---|
| Transport System | | | | | | |
| Rail | n.a. | 4 | 2 | 5 | 1 | n.a. |
| Road | n.a. | 5 | 3 | 4 | 3 | 2 |
| Air | 4 | 5 | 2 | 3 | 4 | 2–4 |
| Sea | n.a. | n.a. | n.a. | 2 | 4 | 4 |

1 = low; 5 = high; n.a.= not affected
Source: Parry and Read (1988)

**Table 12.2 Ranked first-order magnitude of impact on UK Transport: cost per impact**

| Variable | Low cloud | Snow | Rain | Ice | Fog | Wind |
|---|---|---|---|---|---|---|
| Transport System | | | | | | |
| Rail | n.a. | 5 | 5 | 5 | 1 | n.a. |
| Road | n.a. | 5 | 2 | 5 | 5 | 1 |
| Air | 4 | 5 | 1 | 3 | 4 | 2 |
| Sea | n.a. | n.a. | n.a. | 2 | 3 | 5 |

1 = low; 5 = high; n.a.= not affected
Source: Parry and Read (1988)

**Table 12.3 Projected winter temperatures and frost frequencies at selected UK stations based on 1988–89 and 1989–90 data**

| | Deviation of mean temp. 1988–89 | No. of air frosts 1988–89 | Deviation of mean temp. 1989–90 | No. of air frosts 1989–90 | Average number of air frosts 1956–70 |
|---|---|---|---|---|---|
| Plymouth | +2.4 | 4 | +2.4 | 1 | 37 |
| Manchester | +2.7 | 12 | +2.0 | 11 | 42 |
| Aberdeen | +3.3 | 8 | +1.5 | 22 | 60 |

The 1988–89 approximates to likely *mean* temperature changes expected by 2050. By 2030 similar years could be expected to occur in 20% years.

probably the mode of transport most tolerant of adverse weather. For all UK transport systems (except marine), snow is the most serious hazard followed by ice. The relationship between mean winter air temperature and average number of days per month with snow cover in the uplands is suggested in Figure 12.1. Empirical evidence of

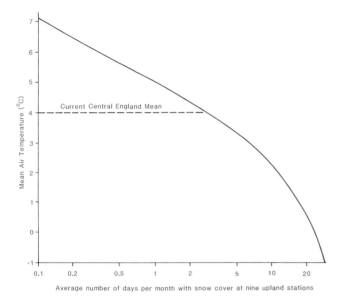

Figure 12.1  Likely changes of days with snow cover in the UK uplands in relation to winter mean temperatures.

weather sensitivity based on content analysis of newspapers and telephone enquiries from the general public to Weather Centres confirms the sensitivity of transport to inclement conditions (Smith, 1989).

*Roads*

A comprehensive analysis of highway meteorology has recently been prepared and includes consideration of the effects of snow/ice, fog and wind on highway planning and operations (Perry and Symons, 1991).

Amongst the beneficial effects on highways of climatic warming would be reduced expenditure on winter maintenance activities. An example of the relationship between winter temperature and winter maintenance expenditure is shown in Figure 12.2. The average annual cost of snow and ice control on Britain's roads is now about £120 million and this has increased as a proportion of the total road maintenance budget. The introduction of new technology for ice detection and the forecasting of minimum road surface temperature has resulted in initial salt savings of at least 20 per cent, and research in Wales based on 'average' winter temperatures, suggests that the cost of such equipment

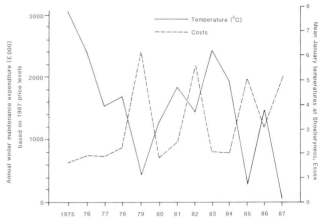

Figure 12.2  Relationship between mean January temperature at Shoeburyness and annual winter maintenance expenditure in Essex, 1975–1987.

can be recouped in as little as two or three years (Perry and Symons, 1991). A predominance of mild winters would greatly extend the pay-back period and is likely to make the deployment of this technology much less of an attractive proposition. However, ice prediction systems could still be expected to contribute to winter maintenance decision making, since milder winters would still include some frost and ice occurrences. Some savings may also be possible in the capital expenditure programme involving snow clearing equipment. Examples of the likely incidence of frost nights based on two levels of projected winter temperature is shown in Table 12.3. Frost, especially when prolonged and severe can result in millions of pounds worth of structural damage to roads and bridges, and this figure could be expected to be reduced. Decreases in the number of freeze-thaw cycles would result in fewer potholes and road damage (Perry and Edwards, 1988).

Climate change would have implications for weather-related road accidents. It has been estimated that the cost of road accidents in 1987 (latest available figure) was in the region of £5 billion, and accidents involving skids caused by bad weather road conditions account for some £300 million (Palutikof, 1991). However, the relationship between bad weather and the number and severity of road accidents is not a simple linear one and less frequent spells of severe winter weather could actually increase accidents since less journeys would be postponed or cancelled. A weak correlation between summer temperatures and accidents reported to the police has been noted over the last few years, probably as a result of less parental control being exerted on children playing outdoors.

Projected increases in precipitation, although small, could exacerbate problems of road flooding, landslips, breaking up of the road surface and corrosion

93

of steelwork on bridges. If storminess increased, disruption from fallen trees and from the overturning of vehicles, such as was widespread in the severe gales of the 25th January and 26th February 1990, could become an increasing problem. Savings in road construction in the projected climate of the 21st century might be expected to include use of thinner surface dressings and less deep sub-grades, although enhanced expansion capabilities could be required and the risk of damage from occasional frosts will remain. Greater incidence of hot weather would mean more frequent adverse weather for resurfacing and surface dressing.

### Railways

Crawford (1989) has noted that 'the basic railway structure presents the opportunity to run a safe and efficient service which although not immune to the vagaries of the weather, can quickly recover from adverse situations'. However, no systematic analysis exists on the likely aggregate impacts on railways of increases in temperature. Snow and ice currently present the most difficult weather-related problem and operating problems and revenue losses from disruption could be expected to fall. At present, punctuality of trains declines when temperature falls below +2°C (Smith 1990). Such events should become less frequent. In addition, investment in technology such as point heaters could be reduced, while potential savings are possible in locomotive rolling stock design and de-icing equipment. An increase in ambient temperature could have an impact on the performance of passenger vehicles with cooling systems designed to current specifications. Increases in flood frequencies, if realised, could weaken and possibly wash away track bed more frequently. If stormy episodes were more frequent, problems could arise on coastal routes (e.g. in South Devon and between Aberdeen and Stonehaven), while bridge failure due to scour (of the type that has occurred at Glanrhyd in Wales and Inverness in recent years) could increase. Changes are possible in rail freight requirements, e.g. less need for movement of bulky commodities like coal to power stations, problems with the movement of crops and foodstuffs sensitive to hot weather.

### Sea and inland waterway transport

Inland waterway transport would be less affected by the occasional freezing of canals and rivers, but perhaps more affected by low water levels as a result of evaporation during hotter summers, when navigation would be difficult.

Increased storminess could affect the viability of high speed light weight passenger craft operation,

perhaps influencing passenger preference in the use of the Channel Tunnel and cross Channel shipping. Marine casualties could be expected to increase during severe stormy episodes and shipping operations in ports and harbours held up. Increases in temperature could require increased use of refrigeration and produce changes in the way cargoes are managed in transit, with implications for the demand for different types of vessels.

### Air transport

Climate warming would be likely to reduce snow and ice problems which now hinder aircraft operations at airports. Air temperature is a critical factor in aircraft performance during take-off, especially at high altitude airports or airports with marginal runway lengths. This factor could be expected to be of little importance at UK airports but could mean reduced loads and higher operational costs for carriers flying into the UK unless suitable aircraft engine power uprating programmes are undertaken. Changes in prevailing winds, were they to occur, could affect the efficiency of runway operations, or re-distribute noise impacts in urban areas. Low cloud amounts could increase, requiring more airports to install precision approach and landing aids to maintain regularity of operations. Changes in the position and intensity of the jet stream could have economic impacts on long-haul flights.

### 12.2.2 Effects of sea level rise

Sea level rises could have an impact on structures such as bridges in some coastal areas, although more damage is likely from the effects of increasing storm surges. Coastal road and rail routes will be vulnerable in low lying areas. In addition, airports located near to sea level, such as those on some Hebridean islands, for example, would be vulnerable to a rise in sea level.

In the case of sea transport, the major impacts of sea level rise could be expected to be negative as a result of damage to port and harbour installation infrastructure. There could be added costs as certain ports become unusable, at least at some stages of the tide.

## 12.3 UNCERTAINTIES AND UNKNOWNS

The potential impacts of climate change in the UK transport sector are not well understood at the present time; information is lacking in the following areas:

- The extent to which some modes of transport, e.g. the private car, might be significantly affected by public policies or consumer actions designed to restrain emissions of greenhouse gases, is unknown. Taxes on carbon based fuels, and stringent fuel economy standards, for example, may cause a shift away from private transport towards greater reliance on public transport. Concern over the build-up in hot weather of low level pollutants may prompt planners to introduce measures such as low speed limits to discourage motorists from undertaking some journeys by private car.

- The extent to which perception of climate change will affect decision-making on weather protection is not certain. Adjustments to, for example, less ice and snow, may be rapid, or the memory of exceptional circumstances from the past may colour judgements in a new climatic environment.

- The influence of meteorological stresses on the incidence of traffic accidents is probably of considerable significance, but requires investigation.

- Information is needed on the extent to which organisations like British Rail, British Airways etc., have conducted in-house research into likely effects of global warming on their operations.

- The distribution amongst transport sectors of the changes in demand for travel, e.g. increased leisure journeys as a result of perceived 'better' summer weather, is unknown.

- Changes in the demand for some goods, such as perishable foodstuffs, coupled with higher ambient temperatures, are likely to affect the pattern and frequency of distribution of goods to wholesalers and retailers to an unknown extent.

## 12.4 PRINCIPAL IMPLICATIONS

Increased temperatures are likely to benefit road systems, requiring less expenditure on maintenance; a decrease in the incidence of rail delays in winter may also occur. However, the volume of road, rail and air traffic may increase as a result of clement weather conditions, which could outweigh these advantages.

Increases in precipitation could exacerbate road flooding, landslips and the corrosion of bridges, resulting in a need for corrective measures. A greater incidence of storms in the UK, such as those of early 1990, may cause increasing problems of

transport disruption as well as considerable damage. Impairment of port facilities and infrastructure could result both from increased storminess and from sea level rise, with concomitant implications for shipping.

## 12.5 RESEARCH AND POLICY NEEDS

### 12.5.1 Future research effort

Comprehensive studies of the likely impacts of climate change on transport have not been carried out and only fragmentary, small scale studies exist. Both broad-scale and specific research is urgently required, for example:

(i) Broad-scale assessments of likely climate-induced demand for transport as life styles, residential and migration patterns change.

(ii) Specific analysis of the level of savings that can be expected in winter maintenance of highways in different parts of the UK. Detailed cost-benefit figures need to be collected to examine whether further investment in costly technology can be justified. A national network of ice detection sensors in a National Ice Prediction System (NIPS) network may no longer be required.

(iii) Specific analysis for engineering design applications of the likelihood of extreme high and low air temperatures in the future, to update the work by Hopkins and Whyte (1975).

### 12.5.2 Policy recommendations

Road transport is a serious contributor to global warming. The present policy of satisfying demands for roads, often at the expense of investment in public transport, needs to be carefully considered, especially in light of the current need for stringent emissions control.

The need for structural storm protection for coastal routes and ports will have to be appraised. Monitoring of bridges and other transport systems may well be required more frequently. Improved road, rail and runway design standards should be considered.

In order to reduce the risks of weather-related accidents, policy effort to sustain research into the design of safer and more weather-resistant forms of transport is also advocated.

# Financial Sector

**SUMMARY**

- There is little information on the potential impacts of climate change upon the various activities within the financial sector, except for the insurance industry, which by the nature of its business would be immediately affected by a shift in the risk of damaging weather events arising from climate change. The cost of severe weather events to insurers and reinsurers has risen steeply (over £1 billion from the October, 1987 'hurricane' alone), with the domestic sector presenting the greatest losses.

- Because of the international nature of financial markets and institutions, the impacts from climate change or sea level rise abroad are likely to be just as important (perhaps even more important) than the impacts in the UK itself.

- If the risk of flooding increases due to sea level rise, this would expose the financial sector to the greatest potential losses (the value of the property protected by the Thames Barrier is £10 to 20 billion).

- If the probability distribution of individual variables such as temperature, rainfall and windspeed changes, then this would alter the frequency of severe weather damage events, and may necessitate pricing or product changes.

- If there were an increased frequency of multiple incidents (e.g. windstorms plus sea surges in the UK or internationally within successive years), this would place financial strain on the insurance industry. The persistent storminess during January and February 1990 led to record losses.

- There will be increasing pressure to improve the quality of information about property exposed to damage. In turn, information on actual weather damage impact will be useful for property owners and design engineers, in order to mitigate or avoid future losses.

- If climate change and sea level rise seem likely to affect socioeconomic activities, this will affect the appraisal of existing and new investment opportunities, particularly in agriculture and for coastal regions, and it could also alter the role of insurance in protecting the most vulnerable areas or activities.

- Financial operations are vulnerable to short-term disruption due to failure of communications or denial of physical access, such as resulted in the October 1987 storm in the London area.

## 13.1 INTRODUCTION AND BACKGROUND

The financial sector of the UK economy generates huge turnover in the UK and in invisible exports, through such services as banking, insurance and broking. The repercussions of climate change could be severe, acting through such media as commodity prices, credit failure and insurance claims. Historically climatic impacts have been viewed as random. Only the insurance industry is beginning to recognise climate change as a strategic variable, both at the UK and international level. Only one company responded to initial enquiries made to the Building Societies Association, banking associations and the Institutional Shareholders' Committee about their strategy for climate change, or indeed their current sensitivity to weather.

This analysis will therefore concentrate on the insurance industry and in particular, the general, or non-life, underwriting sector which is much larger than the UK reinsurance sector and is more sensitive to weather than broking or life insurance. However, it should be recognised that the reinsurance sector plays a major part in coping with weather catastrophies, by spreading the costs more widely. Insurance is essentially concerned with re-distributing the financial risk of human activities. Therefore, although it is useful to consider the direct effect of climate on the insurance industry, the main benefit of this focus is that it provides information about the cost of climate in other sectors, especially domestic consumption. The ancillary aspects of insurance such as risk management and settling claims give insights into the practical problems of damage prevention or recovery.

Most insurance in the UK is provided by free enterprise limited companies, or Lloyd's (the insurers). Those buying insurance (the policyholders) can be subdivided into 'Commercial' (business) and 'Personal' (domestic) policyholders. Although there is a variety of types of insurance cover, including Motor, Marine and Liability, it is *Property* policies which are particularly affected by weather. Usually Personal policies are 'comprehensive', i.e. they cover a wide range of risks, including weather damage, while Commercial policies are more variable; the policyholder may choose to bear part or all of the risk himself and indeed the insurer may exclude or limit certain risks. Businesses can also insure against the consequential loss due to property damage, though this cover is less commonly purchased.

## 13.2 ESTIMATED EFFECTS OF CLIMATE CHANGE AND SEA LEVEL RISE

### 13.2.1 Effects of climate change

Extreme weather causes several types of claim:

Cold — Burst pipes leading to water damage.

Drought — Subsidence of dwellings (not generally covered on Commercial property).

Freshwater Flood — Water damage from rivers/ thunderstorms.

Storm — Wind damage, often associated with water damage.

*Drought*

Long periods with reduced precipitation lead to soil shrinkage, particularly in clay. In turn this can lead to costly claims for *subsidence* to the buildings, or *heave* when heavy rain causes the soil to swell subsequently. Generally only dwellings are covered for this eventuality, not Commercial property. This cover was introduced in 1971, and 'subsidence' claims have amounted to about 10% of the cost of all household claims since 1979.

Figure 13.1 shows a negative relationship between 'annual % damaged of total value at risk' and 'accumulated precipitation' over 1976–89 for a major sector of the UK insurance industry. 1976 was the worst drought since available records began (1727), and 1989 was the 37th – these were the two driest periods during 1976–89. The drought of 1976 occurred not long after the severe wind-storm of January 1976 without causing long-lasting harm to the insurance industry. If precipitation were to

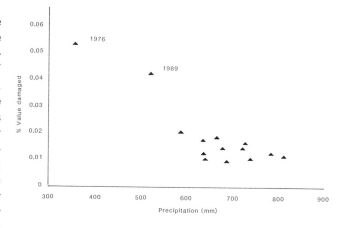

Figure 13.1 Annual damage to buildings *versus* accumulated rainfall, December–August (1976–1989). (Source: Climate Monitor and unpublished insurance industry statistics).

*increase* by about 90mm each year for the period December to August (as suggested in Section 2), this would tend to reduce subsidence claims, but the effect might be more than offset by swifter evaporation due to increased temperatures.

Subsidence is localised, therefore the relief will be scattered. Though costly, claims are not numerous (less than 1 per 1,000 dwellings per year), so there will not be any significant change to insurers' operations if the volume of claims is reduced. One industry estimate gives a cost per year of £220M at 1990 values during the period 1984–89 (Shearn, 1990).

### Freshwater flood

Such events arise from local thunderstorms, or from prolonged periods of precipitation, possibly exacerbated by snow-melt and coastal tides. Data on frequency and cost are not readily available, due to this being a relatively minor hazard. It accounts for perhaps 2.0% of the total cost of household claims annually, with no major deviations since 1976 when data became available. Data are less complete for Commercial business, but the total insured cost annually is probably similar to the total household cost, about £25M (1990 values).

If mean annual precipitation were to increase by about 10% (the average annual value for England and Wales for 1979–88 was 928mm), this could result in noticeable additional cost, depending on the new pattern of precipitation. Nevertheless the effect is not seen as critical.

### Summer temperatures

Property is relatively insensitive to higher temperatures (within UK parameters), so increased temperature is not likely to increase insurance claims. There is a possibility of heath or woodland fires, but this is not seen as a serious threat.

### Winter temperatures

The main threat to Property from low temperatures is the bursting of waterpipes, whether from central heating, waste, domestic, industrial, or sprinkler systems. The major cost arises not from the repairs to the damaged systems, but from spoilage of stock, fittings and furniture. The rise in consumer wealth has made this a particularly costly hazard in domestic property insurance. Typically the damage arises in a matter of days, and statistics are not generally available in such detail. Also, 'accidental' claims are not generally separated from 'weather' claims. Preliminary analysis suggests that as much as 50% of 'burst pipe' costs are not due to winter weather. Relative to storm claims, burst pipe claims are about twice as expensive each, but are far less numerous. Nevertheless, difficulty may arise in the recovery process due to shortage of skilled labour (plumbers etc.).

At 1990 values, the cost over the period 1984–89 of weather damage to water systems within property is estimated at £125M annually. Considering domestic claims alone, the projected rises in winter temperature will not eliminate 'winter' burst pipe claims, but could substantially reduce them: perhaps by as much as 25% for the temperature increase projected for 2010 (+0.7°C); by 30% for the temperature increase projected for 2030 (+1.4°C); and, by 35% for the temperature increase projected for 2050 (+2.1°C).

However, these impacts clearly depend on whether the distribution of temperatures around the mean remains similar to that prevailing today or whether it is substantially altered. Little is known about this at present.

### Windstorm

This hazard is a major component of insurance costs. It also causes operational problems due to the high volumes of claims handled in the aftermath, and the difficulty of obtaining skilled tradesmen to carry out repairs. Policy cover is usually restrictive in trying to reduce the inclusion of 'maintenance' repairs, particularly for boundaries to property and removal of trees. Unfortunately the scenarios give no assistance with future trends for this variable. The insurance industry does feel that such incidents are increasing, with major events in January 1976, January 1984, October 1987, January 1990 and February 1990 in the UK, and further incidents in Europe during that period. Storms accompanied by rain are more serious due to the water damage to exposed interiors. Tree-fall is a significant cause of damage to property and disruption to transport and communications.

Figure 13.2 shows that the proportion of value insured which was damaged in the areas of south east England affected by '87J' (the October 1987 windstorm) varied with windspeed. It is notable that the highest speeds recorded were about 48 m/s, or 108 mph. At that speed, the proportion damaged rose to about 0.7% of the value insured (ignoring interiors completely). The proportion by number was much higher in the affected areas, at over 40%; 75% of local authority housing was affected in some areas (Buller, 1988). In tropical cyclones, values damaged can reach as high as 30.0% of the sums insured. Domestic property is more susceptible to storm damage, as indicated by available statistics. (About 30% of the insured cost of storms is Commercial, 70% Domestic). The cost of individual major events has exceeded £1,000M in 1990 values, and this is certainly causing some concern.

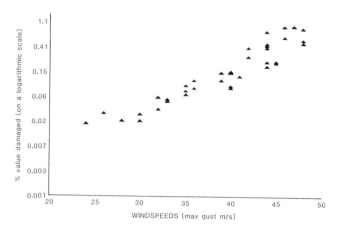

Figure 13.2 Cost of damage *versus* windspeed. (Source: Buller, 1988 and unpublished insurance industry statistics).

Windstorm can also cause sea flooding by increasing wave height and generating a storm surge (see 13.2.2).

## Non-property insurance and climate

Motor claims are sensitive to fog and road-ice (as well as precipitation). Heat, atmospheric pollution and increased levels of ultra-violet radiation may be important for health and life insurance. Clearly, large-scale loss of life in a natural catastrophe would have implications for life-insurers.

Liabilities insurance is unlikely to be affected, due to the difficulty of separating negligence from natural causes, although there is some concern at the risk of pollution from waste sites following large-scale inundation. The question of public authorities' liability to protect the community from natural hazards might arise if protections or warnings are inadequate. Liability for damage caused by tree roots during drought is already recoverable in law.

### 13.2.2   Effects of sea level rise

Flood insurance has been generally available to householders and small traders since 1961. There is no doubt that property could be severely affected by sea inundation. Since 1953 there has not been a major incident, and minor incidents are not coded separately from storms or floods. The 1953 floods cost between £240 million and £400 million in 1982 values, of which only £52 million was insured (Arnell, 1984). A rise in eustatic water level could be exacerbated by severe storms, with associated higher wave height and wind surge.

### 13.3   UNCERTAINTIES AND UNKNOWNS

## Insurance

Better information on future precipitation patterns will allow a firmer assessment of likely changes in subsidence and freshwater flood claims. Winter warming patterns may reduce burst pipe claims, but this depends on the exact patterns observed. However, Figure 13.3 suggests that any saving in burst pipe claims would be outstripped by the increase in storm claims. A major unknown is the future wind-circulation pattern, which has serious possibilities for property damage. Hail is a peril which proved very costly in Munich in 1984 and could affect the UK (Dimmock, 1989). Also, by 2030 it is possible that heath and woodland fires might become a major hazard in dry years due to the higher average temperatures likely to prevail. Better information is required on the sensitivity of property to weather damage – historical statistics are difficult to assemble. As regards sea level changes, one of the major unknowns is the value and location of property most at risk. The exposure to flood in London was estimated at between £10 billion and £20 billion, justifying the construction of the Thames Barrier, but in general exposures have not been assessed (Hitchcock, 1989). Transport, life and health insurance are all likely to be affected by climate change, but no estimates can be given.

An additional concern is the chance combination of events, which can yield exceptional extremes, as happened at Towyn where the storm surge, wave height and water level were individually not exceptional, but when combined had a return period of 1000 years (Roe, 1990).

## Other financial industries

The operation of credit institutions could be affected indirectly by natural catastrophes. The need for cash to pay for property damage could reduce customers' balances significantly. 'Red-lining' by insurers, or the abandonment of property, could render worthless the assets used to secure loans. While these possibilities are now recognised, their likelihood is not yet quantified.

### 13.4   PRINCIPAL IMPLICATIONS

## General insurance

Figure 13.3 gives an illustration of the impact on Property insurance of severe weather since 1960, at 1987 values. The cost for 1990 is a very early estimate. The figures present an increasing trend, but they are affected by (i) increasing real exposure

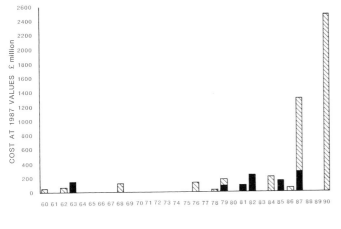

Figure 13.3 Cost of major U.K. weather incidents, 1960–1990.
(Source: Association of British Insurers; 1990 data
from press).

during the period, due to growth in population, wealth, wider insurance cover, etc., (ii) greater propensity to lodge claims and (iii) a return to 'normal' variability following the very uneventful 1960s. From a study of non-catastrophe costs, the effects of (i) and (ii) during the period 1980–89 were to double the impact of weather damage. It can be seen that the costs during this period have in real terms escalated much faster than that. Recent storms have proved far more costly than the cold winters of 1981/2 and 1987.

To understand the figures, note that they exclude: subsidence and 'normal' weather; transport insurance, personal injury, consequential loss; uninsured or unclaimed losses. They are not completely comprehensive (e.g. March 1987 windstorm is ignored) and sometimes aggregate distinct incidents (October, 1987 storms and floods). They do not allow for differences in exposure between the impacted areas; and, generally, only one fairly early estimate of cost is given with no subsequent update (however, October, 1987 was updated due to the slowness in claims being reported). It is notable that windstorm has now become a much greater peril than burst pipes.

The mix of building stock changes only slowly, and with current structures, critical thresholds for damage appear to be as follows:

Drought — less than 500 mm precipitation over England and Wales in the period December–August

Burst Pipes — temperatures of less than −5°C for more than 24 hours (Dlugolecki, 1989).

Windstorm — gusts of 90 mph.

Flood — thresholds for freshwater flooding depend very much on local circumstances (Hempsell, 1961). Seawater flooding is estimated to be 10% more costly than freshwater and, in general, inundation for more than 12 hours can add more than 25% to the cost of damage (Davies, 1983). Furthermore, water contaminated by sewage necessitates very costly repairs, (between £8,000 and £36,000 per house) (Roe 1990).

More extreme weather than that described above will cause significant damage, and may be of national importance, depending on the values exposed. Locally, damage will reflect topography, soil structure, building structures, etc. For example, damage at Towyn was more costly because most of the stock were bungalows, with greater vulnerability to ground water damage (Roe, 1990).

Insurance has coped with the past rate of change by using pricing, minimum damage limits (deductibles or excesses), and risk-spreading through reinsurance to alleviate the financial effects. Risk management has been tackled through distribution of leaflets and publicity and physical inspection of individual premises followed by recommendations for physical protection, often with some adaptation of the cover given. The operational effects of catastrophic storms have been alleviated by easier and faster guidance to claimants on what to do. This has enabled the industry to cope with a doubling of costs in a decade. The 1990 events represent a *further* doubling, and may be expected to throw some strain on the profitability of the industry. The signs are that, operationally, the industry is coping, having learned valuable lessons from the October 1987 storms. However, competitive pressures may limit the application of some of the traditional remedies, e.g. price increases.

The financial strain is likely to be exacerbated by multiple incidents occurring either in the UK or internationally, placing a great need for substantial capital and reserves. While insurance is linked to other sectors through their weather losses, it is the domestic sector which generates the greatest losses owing to its size, vulnerability and width of cover freely available.

*Other financial industries*

The principal implications of climate change for other financial industries are summarised below:

● The basic operation of credit is underpinned by insurance in many instances, and thus is dependent on the solvency of the insurance industry and its willingness to continue to offer cover.

101

- Failure of telecommunications/electric power for a prolonged period (i.e. a day) would be serious to specific operations – for example, on October 16, 1987, when international markets crashed due to global trading factors, at the same time the UK was hit by a severe windstorm which prevented the UK markets from trading.
- The international nature of financial markets means that climatic impacts in other countries could quickly affect the UK financial sector.

## 13.5 RESEARCH AND POLICY NEEDS

### 13.5.1 Future research effort

*Insurance*

(i) Predictions of future probability distributions of temperature, rainfall and windspeeds are required to allow assessment of the frequency and severity of different losses, including the possibility of combinations of events.

(ii) More precise quantification of the links between different levels of meteorological event and physical and financial damage is needed.

(iii) Information on the exposed risk must be improved (particularly the locations of individual buildings insured under 'block' building society accounts), and, in general, on vulnerability to flood.

(iv) Effective ways to encourage risk management by policyholders need to be established.

*Other financial industries*

(i) Resources should be allocated to explore climate change as a variable at UK and international level.

(ii) Investments must be classified according to climatic vulnerability.

### 13.5.2 Policy recommendations

*Insurance*

(i) It is vital that insurers have adequate reserves to meet catastrophe payments. At present UK taxation policy imposes a penalty on 'catastrophe funds', contrary to German practice. With the advent of the Common Market, a 'level playing field' must be created. The alternatives are reinsurance (which will never be unlimited in supply, may increase sharply in price when it is required most and may transfer profits abroad) or Government assistance

(which is contrary to current policy that favours free enterprise, and which imposes a burden on the tax payer).

(ii) Owners should physically protect and maintain their assets to bring the risk of damage within insurable parameters. Expert advice should be made available and insurance terms adapted to individual risk circumstances. However, for domestic property this may not be economic other than by remote contact, and the scope for varying terms for flood is limited by public policy (Arnell, 1984). Also, in practical terms, the proportion of premium related to any one hazard is relatively small thus reducing the scope to encourage risk protection on most property. In view of the general size of premiums, discounts would not give sufficient encouragement to invest in protection and there is considerable customer inertia. Where the hazard is obvious, e.g. lagging water pipes, protection should be automatic anyway. However, recent trends in consumerism make it unlikely that an insurer could succeed with the argument that the policyholder showed lack of reasonable care.

(iii) Public emergency procedures should be improved to alleviate losses when severe weather is imminent, e.g. more effective public warning systems.

(iv) The 'recovery' phase should incorporate prompt advice and reliable repairs properly controlled. Such services are increasingly provided by insurers and remove a considerable burden from the local authority.

(v) Insurers should provide information on damage patterns to technical design centres and others to ensure adequate attention to climatic factors and inundation, and to assist cost-benefit studies, when setting standards for the design and maintenance of buildings or coastal defences.

*Other financial industries*

(i) Existing investments and strategies should be reviewed in the light of probable climate change and sea level rise, particularly with respect to agriculture and coastal regions.

(ii) In the appraisal of new projects, climate change and sea level rise must be considered.

(iii) Personal and business assets should be able to purchase insurance cover during their lifetime in order to provide a solid basis for credit. Particularly vulnerable property may require other financial mechanisms to spread losses and protect poorer members of society from financial hardship. (In the recent Towyn floods 6% of the affected buildings and 38% of contents were uninsured).

(iv) Emergency procedures should be improved to ensure that operations can continue with minimum disruption following natural catastrophes.

# Recreation and Tourism 14

**SUMMARY**

- Tourism in the UK has an international dimension which is sensitive to any change in climate that alters the competitive balance of holiday destinations worldwide.

- Increases in temperatures are likely to stimulate an overall increase in tourism in the UK, with the greatest effects on activity holidays and some forms of outdoor recreation.

- If any change towards warmer, drier summer conditions occurs there may need to be more restricted public access to large areas of upland Britain because of the enhanced risk of fire.

- Increases in temperature in the UK will have an effect in upland Britain, where the balance between outdoor recreation and agriculture may be altered, if, for example, more intensive agriculture reduces public access to the countryside.

- The commercial viability of some winter sport developments in upland Scotland would be at risk if snowfall and snowcover are reduced.

- Any significant increases in rainfall, windspeed or cloud cover could offset some of the general advantages expected from higher temperatures.

- Rises in sea level would affect fixed waterfront facilities such as marinas and piers. For beaches backed by sea walls, increased erosion would lead to a lowering of the beach and subsequent undermining of the walls. Recreational habitats such as sand dunes, shingle banks, marshlands and soft earth cliffs would also be affected.

- An increase in sea temperatures would increase the pressures of tourism on UK beaches, while coastal erosion may reduce beach area.

## 14.1 INTRODUCTION AND BACKGROUND

Tourism and outdoor recreation is one of the most important and rapidly growing service industries throughout the world. Within the UK, tourism is a major source of employment. Although people travel for many reasons it is likely that countries whose economies are highly dependent upon tourism will face the greatest challenges from climate change.

The continued success of tourism is intimately associated with the quality of the renewable biophysical resources on which it rests. The state of these resources is climate-dependent. In some parts of the world, climate itself is the resource which has directly promoted tourism. Despite this, the leisure industry has so far shown little interest in exploring either its current atmospheric sensitivity or the possible impact of future climate change. In the UK the variety of the landscape and the rich cultural heritage are major assets. Any adverse disturbance of this pattern and reduction of the tourist experience is likely to have major economic implications for the tourist industry.

In the absence of reliable model-based predictions of climate change at a regional scale, any consideration of the effects of the greenhouse-gas induced warming on tourism is highly speculative. Tourism in the UK, as in other mid-latitude areas, will experience a complex mix of winners and losers. For example, the conclusion from a series of studies undertaken in Ontario, Canada, has been that, although the ski season would be reduced, the opportunities for summer recreation will be enhanced by a longer season *provided that* the predicted decline of water levels does not adversely affect wetland-based activities (McBoyle, *et al.,* 1986). In addition, it is important to note that UK tourism will be affected by global, as well as by regional and national, changes in climate and tourist opportunities.

## 14.2 ESTIMATED EFFECTS OF CLIMATE CHANGE AND SEA LEVEL RISE

Climate change and sea level rise will separately affect the volume of tourism, the type of tourism, and the safety of some outdoor recreational activities.

### 14.2.1. Effects of climate change

*Volume of tourism (domestic and overseas)*

Climate influences the total extent of tourist activity in complex ways. For example, the volume of UK summer tourist visits to the Mediterranean is sensitive to seasonal climatic variations in Britain. Statistical analysis, significant at the 1% level, has demonstrated that such vacation travel decisions are influenced by precipitation conditions in the UK during the preceding summer (Smith, 1990).

Climate change resulting in increased temperatures alone would increase the total volume of tourism in the UK by raising both day trip and stay trip activity. In the much warmer winters indicated by the models, many winter vacations currently taken in the Mediterranean or further afield by British residents could well become less compelling. This also applies to long-stay winter visits overseas, which is relevant in the context of the growth of the 'silver market' (people age 55+ and early active retired with a substantial disposable income). During the summer, some Mediterranean locations may become unpleasantly hot. Conversely, any trend towards warmer UK summers, especially if it significantly improved the temperature of bathing waters and was not accompanied by higher rainfall, would encourage more UK domestic tourism and attract additional overseas visitors.

In the global context, Britain may also gain from the relative decline of other overseas destinations. Many developing countries in the Caribbean, Africa and the Far East are increasingly dependent on tourism based on warm climates and attractive beaches. These destinations include several groups of small, low-lying tropical islands which are especially vulnerable to rising sea levels. Less developed countries may well lack the resources to combat rising sea levels and be unable to offer other facilities. Several tropical beach-based destinations may face additional hazards, such as an increase in cyclone activity and storm surges resulting from higher sea surface temperatures.

On the other hand, much UK tourism is probably less climate sensitive. Visits to friends or relatives, enjoyment of the scenery and 'cultural tourism' (visits to historic cities and heritage sites) are less likely to be affected by climate change. Higher summer temperatures will increase the potential for some mainstream outdoor activities like swimming, beach use and camping. Other activity holidays like golf, tennis, visiting parks and other outdoor sites may start to take place more during the 'shoulder' months of spring and autumn. Activity holidays already account for almost one-quarter of all UK holiday expenditure. Any such trend towards increasing this market and lengthening the tourist season would be welcomed by the industry. If summers become hotter and drier, more tourist activity will impose a further seasonal peak demand on water supplies.

However, these assumptions are conditional on the overall 'quality' of the weather that the visitor experiences. This is dependent on a much wider range of atmospheric conditions than temperature alone. Future precipitation regimes, as well as some of the more 'aesthetic' elements of recreation climatology, like cloudiness and windiness, will be crucial as tourists become more discriminating. Surveys conducted by the Scottish Tourist Board reveal that the biggest single cause of visitor dissatisfaction in Scotland is the weather (Hay, 1989). Typically over 10% of British, and nearly 20% of overseas visitors complain about the weather.

Increased temperatures alone should increase the demand for UK tourism as some long holidays abroad are partially substituted by further growth in the domestic short-break market. But any trend to wetter conditions, including more wind, cloud and reduced visibility, would probably not be beneficial. At the very least it would entrench investment in wet weather facilities and would further invalidate the undue emphasis on promotional brochures full of 'blue sky' illustrations. Warmer summer conditions may be generally welcome but there are many possible side effects. For example, changes in humidity/wind/temperature relationships may increase the concentrations of biting insects in the peak summer months. The microclimates of cities are likely to become much less pleasant. This may force consumers out of the urban areas, where they enjoy mainly cultural activities, to the cooler resorts and the fringes of the cities to consume more rural leisure-related pursuits. To some extent the attractiveness of the traditional beach resort may be reduced as a result of concern about the exposure of sensitive skin to increased UV radiation.

*Type of tourism*

Climate change is likely to alter the types of holiday activity pursued by tourists in the UK. The most fundamental effects concern the presence, or otherwise, of 'direct' climatic resources such as snow cover for winter sports or water temperatures sufficiently high for bathing. Low level ski resorts, occupying small areas of low mountains in Scotland, are especially sensitive to variations in winter temperature and snowfall. The warm winters of 1988–89 and 1989–90 produced a diversification into other activities, such as hill walking and the use of indoor facilities. With the possible increase in the frequency of such winters, this could prove an early indication of the longer-term adaptation which will be required.

The latitude and altitude of recreation areas will largely determine the relevance to them of climate change. Upland areas are complex climatic environments with steep gradients of weather and climate over short distances. In northern Britain the upland extension of agriculture, and other land use changes associated with a longer thermal growing season, could well bring about an increased demand for land uses which would conflict with outdoor recreation. For example, more suitable growing conditions for trees in the uplands and more intensive animal grazing could bring pressure for reducing public access to parts of the countryside. Shifts of biotic types within National Parks could affect their tourist potential.

Climate changes could affect many other ecosystems on which outdoor recreation depends. The 1976 drought in England and Wales, hailed by some commentators as a taste of things to come, created severe problems for water, land and wildlife management. Any persistent change to warmer drier conditions in summer, with an increased risk of forest and heathland fires, may well result in large areas of open country in the National Parks being closed to summer visitors.

A large amount of outdoor summer recreation takes place along the shorelines of freshwater lakes and reservoirs. Increases in temperature, plus associated changes to the hydrological regime, may change the tourist potential of such shorelines. For example, any fixed waterfront facilities, such as marinas, will be vulnerable to either positive or negative changes in water level. Tourist potential (particularly the emerging demand for 'green' tourism) may decline if some wetlands disappear with a consequent loss of ecological diversity. Colder water species of game fish may be replaced by other varieties and reduce the attraction of fishing, which is reputedly the main participant sport in Britain.

Participation in water sports like sailing has shown a marked growth in recent years. Any decrease in summer river flows, combined with higher temperatures, could result in higher levels of water pollution. A less attractive recreational environment would arise not only from the reduced dilution offered to effluent in lakes and rivers but also due to the warmer water temperatures, which will encourage higher rates of production of algal blooms and bacteria. Any trend to wetter conditions, combined with further improvements in the drainage of agricultural land, would create more spates on rivers and increase sediment yields. The effect would be to reduce the recreational potential of freshwater fisheries.

*Safety of outdoor recreation*

For some outdoor sports and activities, the level of safety experienced by participants is crucially dependent on the weather. Warmer winter conditions will bring less snow but more risk of snow avalanches when snow has fallen. Avalanches are already a growing hazard in many mountainous winter sports areas in Europe and North America (Smith, 1988). This may result in more frequent and costly closure of ski slopes, at a time when the general opportunities for skiing are being reduced, with greater controls on other mountain activities like climbing and backpacking. Any trend towards increased storminess will directly affect the safety of pursuits such as small boat sailing and gliding.

### 14.2.2 Effects of sea level rise

Rising sea levels are likely to have profound effects for recreation along all marine shorelines. Boorman, et al. (1989) have documented the anticipated changes to the British coast resulting from rising sea levels. For a beach backed by a sea wall, as in many resort towns, the prediction is that increased erosion would lead to a lowering of the beach. Depending on the supply of sand, the beach facility could be totally lost with subsequent undermining of the stability of the sea wall. The costs of shoreline protection, plus any beach preservation and replenishment, will be high.

Other UK coastal habitats used for recreation such as sand dunes, shingle banks, marshlands and soft earth cliffs would also be affected, as will any recreational facilities built along the shore such as swimming pools, recreational areas, etc.

But these effects may not be worse than impacts elsewhere. The low gradient beaches characteristic of much of the Atlantic and Gulf Coasts of the USA are very vulnerable to erosion (Leatherman, 1989). In other countries important historic sites, such as the shoreline temples of Bali and Tamil Nadu (India) and the city of Venice (Italy), are under threat from rising sea levels and storm surge.

### 14.3 UNCERTAINTIES AND UNKNOWNS

The weakness of many climate impact studies is that they assume that 'everthing else will remain equal' over the next few decades. Tourism and recreation will react especially strongly to the economic, social and technological changes which will take place in the future.

The future demand for tourism in the UK (especially the domestic demand) will depend on the availability of both increased leisure time and increased disposable income. Of these two, disposable income is the more important. At the present time,

average disposable income is growing more rapidly in many overseas countries, such as the USA and around the Pacific Rim, than in the UK.

Given the availability of high disposable incomes, technology can already go some way to countering the effects of climate and climate change. For example, totally weather-proofed indoor facilities (water-based and theme-based) already exist in the UK but, if general conditions for recreation improve, there may be less demand for these. More specialised weather-proof facilities, such as an indoor ski resort in the Tokyo Bay area of Japan, are also technically possible.

Investment in high-cost developments will be linked to the expected financial return. Unless UK disposable incomes rise further in real terms, the unit market price is unlikely to be sufficient to attract the necessary investment in these new facilities. Multinational corporations may switch investment funds away from tourist developments which are perceived to be climatically vulnerable, both in the UK and elsewhere. These economic factors are highly unpredictable.

More extended and more accurate weather forecast information from agencies such as the Meteorological Office will be an important element in allowing the tourist to optimise on future weather sensitivity. The issue of reasonably accurate 5-day to 30-day forecasts, for example, could stimulate further growth in 'spontaneous' short-break holidays, particularly in the shoulder months. Any trend towards extended weather forecasting is likely to encourage more tourists to take advantage of predicted spells of fine weather and thereby increase congestion and management problems for the industry at 'honey pot' sites.

### 14.4 PRINCIPAL IMPLICATIONS

The following recreational environments in the UK are particularly sensitive to future climate change:
- Winter snow conditions in the Scottish mountains are critical for the future of downhill skiing.
- Spring, summer and autumn conditions in the British uplands are critical for determining the competitive land use balance between outdoor recreation and agriculture.
- Summer water temperatures and other water quality parameters are critical for most water sports and activities.
- Sea level rises are critical for beach holidays and other coastal ecosystems which attract tourists.

## 14.5 RESEARCH AND POLICY NEEDS

### 14.5.1 Future research effort

There is a general need to communicate the possible global impacts of climate change to policy makers in tourism and to relate the possible climate changes to the planning horizons within the industry. At present, the tourist industry generally does not look more than 10 years ahead (a 5-year planning period with another 3–4 year implementation period).

There is insufficient understanding of the current weather and climatic sensitivity of tourism in the UK. Further research is essential before the impacts of climate change can be properly addressed. Baseline studies of recreation climatology, similar to those available in Canada, Australia and elsewhere, are required (Paul, 1972; Masterton and McNichol, 1981; de Freitas, 1990). These need to be supplemented by behavioural surveys. Studies of weather-related holiday decisions, activity patterns and visitor responses to varying weather conditions at key tourist sites will require some painstaking basic research.

The most urgent need is to determine the effect of the phased increases of winter temperature currently projected by climate models on snowfall and snowcover in Scotland in order to assess the future viability of downhill skiing. These temperature changes may be accompanied by changed airmass patterns leading (as in very recent years) to significant differences in the snow accumulation between resorts in the east and west of the country.

For UK tourism, any increases in rainfall may be more significant than temperature increases. At the very least, higher rainfall would offset some of the general advantages expected from higher temperatures. Ideally, there is a need to provide future climate scenarios which address rainfall conditions and also include information about other climatic parameters relevant to tourism.

### 14.5.2 Policy recommendations

There is little doubt that the climate should feature more prominently in the minds of planners and policy makers who have responsibility for tourism and recreation. A greater awareness of climatic variations is needed not only for long-term planning but also to deal with existing conditions.

The tourist industry has much to learn about the practical application of weather and climate information. As the tourists become more sophisticated, and as awareness of climate and climate change grows, resort areas will have to improve their marketing. Some campaigns might attempt to reduce the seasonality of bed nights and encourage more visitors in the shoulder months when the weather is often suitable (and may possibly alter for the better) for a range of outdoor activities.

The projected winter warming over the higher latitudes of the UK poses a distinct threat to the viability of the Scottish ski industry. Bearing in mind the lengthy planning and lead-in time necessary for such developments, the implications of climate change should be reflected in any future policy decisions. The scope for recreational flexibility and possible diversification into other activities could become an important part of such strategy.

In view of the likely important changes to ecosystems and resulting competition for land in the British uplands, the Government should invest in more high level climate stations so that the output from the process-based models, and the actual changes in climate, can be checked and validated for upland areas.

# References

*Section 1*

Department of the Environment (1988) Possible Impacts of Climate Change on the Natural Environment in the United Kingdom. Department of the Environment, London.

Parry, M. L. and Read, N. J. (1988) The Impact of Climatic Variability on UK Industry. AIR Report 1, Atmospheric Impacts Research Group, University of Birmingham, Birmingham.

Pearce, D., Markandya A and Barbier, E. (1989) Blueprint for a Green Economy. Report to the UK Department of the Environment. Earthscan, London.

*Section 2*

Bretherton, F., Bryan, K. and Woods, J. D. (1990) Time-dependent greenhouse-gas-induced climate change. In: Houghton, J. T., Jenkins, G. J. and Ephraums, J. J. (eds.), Climate Change: The IPCC Scientific Assessment. Cambridge University Press, Cambridge, 173–194.

Dickinson, R. E. (1986) How will climate change? In: Bolin, B., Döös, B. R., Jaeger, J. and Warrick, R. A. (eds.) The Greenhouse Effect, Climate Change and Ecosystems. John Wiley and Sons, Chichester.

Hansen, J., Lacis A., Rind, D., Russell, G., Stone, P., Fung, I., Ruedy, R. and Lerner, J. (1984) Climate sensitivity: analysis of feedback mechanisms. In: Hansen, J. and Takahashi, T. (eds.) Climate Processes and Climate Sensitivity. Geophysical Monograph 29, Maurice Ewing Vol. 5, American Geophysical Union, Washington D.C., USA.

Hoffert, I. M. and Flannery, B. P. (1985) Model projections of the time-dependent response to increasing carbon dioxide. In: MacCracken, M. C. and Luther, F. M. (eds.), Projecting the Climatic Effects of Increasing Carbon Dioxide. DOE/ER-0237, U.S. Department of Energy, Carbon Dioxide Research Division, Washington, 149–190.

Houghton, J. T., Jenkins, G. J. and Ephraums, J. J. (eds.) (1990), Climate Change: The IPCC Scientific Assessment. Cambridge University Press, Cambridge.

Hulme, M. and Jones, P. D. (1989) Climatic change scenarios for the UK. Report for the Institute of Hydrology, Contract No. F3CRO5-C1-31-01.

Jones, D. E. (1987) Daily Central England temperature: recently constructed series. Weather, 42, 130–133.

Manley, G. (1974) Central England temperature: monthly means 1659–1973. Q. J. Royal Met. Soc., 100, 389–405.

Mearns, L. O., Katz, R. W., and Schneider, S. H. (1984) Extreme high temperature events: Changes in their probabilities with changes in mean temperature. J. Clim. Appl. Meteorol., 23, 1601–1613.

Mitchell, J. F. B., Manabe, S., Tokioka, T. and Meleshko V. (1990) In: Houghton, J. T., Jenkins, G. J. and Ephraums, J. J. (eds.), Climate Change: The IPCC Scientific Assessment. Cambridge University Press, Cambridge, 131–164.

Oerlemans, J. (1989) A projection of future sea level. Climatic Change, 15, 151–174.

Parry, M. L. (1985) The impact of climatic variations on agricultural margins. In: Kates, R. W. with Ausubel, J. H. and Berberian, M. (eds.) Climate Impact Assessment: Studies of the Interaction of Climate and Society. SCOPE 27, John Wiley and Sons, Chichester, 351–367.

Raper, S. C. B., Warrick, R. A. and Wigley, T. M. L. (1991) Global sea level rise: past and future. In: Milliman, J. D. (ed.), Proceedings of the SCOPE Workshop on Rising Sea Level and Subsiding Coastal Areas. Bangkok, 1988. John Wiley and Sons, Chichester.

Santer, B. D., Wigley, T. M. L., Schlesinger, M. E. and Mitchell, J. F. B. (1990) Developing climate scenarios from equilibrium GCM results. Max Planck Institute fur Meteorologie Report No. 47, Hamburg, FRG.

Schlesinger, M. E. and Zhao, Z-C. (1989) Seasonal climate changes induced by doubled $CO_2$ as simulated by the OSU atmospheric GCM mixed-layer ocean model. J. Climate, 2, 459–495.

Stouffer, R. J., Manabe, S. and Bryan, K. (1989) Interhemispheric asymmetry in climate response to a gradual increase of atmospheric $CO_2$. Nature, 342, 660–662.

Warrick, R. A. and Oerlemans, J. (1990) Sea level rise. In: Houghton, J. T., Jenkins, G. J. and Ephraums, J. J. (eds.), Climate Change: The IPCC Scientific Assessment, Cambridge University Press, Cambridge, 257–281.

Warrick, R. A. and Farmer, G. (1990) The greenhouse effect, climate change and rising sea level: implications for development. Trans. Inst. Br. Geogr., 15, 5–20.

Washington, W. M. and Meehl, G. A. (1984) Seasonal cycle experiment on the climate sensitivity due to a doubling of $CO_2$ with an atmospheric general circulation model coupled to a simple mixed-layer ocean model. J. Geophys. Res., 89, 9475–9503.

Washington, W. M. and Meehl, G. A. (1989) Climate sensitivity due to increased $CO_2$: experiments with a coupled atmosphere and ocean general circulation model. Clim. Dyn., 4, 1–38.

Wetherald, R. T. and Manabe, S. (1986) An investigation of cloud cover in response to thermal forcing. Climate Change, 8, 5–23.

Wigley, T. M. L. (1985) Impact of extreme events. Nature, 316, 106–107.

Wigley, T. M. L. and Jones, P. D. (1987) England and Wales precipitation: a discussion of recent changes in variability and an update to 1985. J. Climatology, 7, 231–246.

Wigley, T. M. L. and Raper, S. C. B. (1987) Thermal expansion of sea water associated with global warming. Nature, 330, 127–131.

Wigley, T. M. L., Santer, B. D., Schlesinger, M. E. and Mitchell, J. F. B. (1990) Developing scenarios from equilibrium GCM results. Climatic Change, (submitted).

Wilson, C. A. and Mitchell, J. F. B., (1987) A doubled $CO_2$ climate sensitivity experiment with a global climate model including a simple ocean. Journal of Geophysical Research, 92 (13), 315–343.

*Section 3*

Boden, J. B. and Driscoll, R. M. C. (1987) House foundations – a review of the effect of clay soil volume on design and performance. Municipal Engineering, 4, 181–213.

Boorman, L. A., Goss-Custard, J. D. and McGrorty, S. (1989) Climatic change, rising sea-level and the British coast. Institute of Terrestrial Ecology Research Publ. No. 1, HMSO, London.

Bouwman, A. F. (1990) Exchange of greenhouse gases between terrestrial ecosystems and the atmosphere. In: Bouwman, A. F. (ed.), Soils and the Greenhouse Effect. John Wiley & Sons, Chichester, 61–128.

Building Research Establishment (1990) Assessment of damage in low rise buildings. BRE Digest 251, London.

Burke, M. K., Houghton, R. A. and Woodwell, G. M. (1990) Progress towards predicting the potential for increased emissions of $CH_4$ from wetlands as a consequence of global warming. In: Bouwman, A. F. (ed.), Soils and the Greenhouse Effect. John Wiley & Sons, Chichester, 451–456.

Hazelden, J., Loveland, P. J. and Sturdy, R. G. (1986) Saline soils in North Kent. Soil Survey Special Survey, No. 14. Soil Survey of England and Wales, Harpenden.

Ineson, P. (1990) Personal communication.

Jenkinson, D. S., Hart, P. B. S., Rayner, J. H. and Parry, L. C. (1987) Modelling and turnover of organic matter in long term experiments at Rothamsted. Intecol Bulletin 15, 128.

Jenkinson, D. S. (In press). The turnover of organic carbon and nitrogen in soil. Phil. Trans. Royal Soc. Lond. B.

Jones, R. J. A. and Thomasson, A. J. (1985) An Agroclimatic Databank for England and Wales. Soil Survey Technical Monograph No. 16, Harpenden.

Jones, R. J. A. and Thomasson, A. J. (1987) Land suitability classification for temperate arable crops. In: Beek, K. J., Burrough, P. A. and McCormack, D. E. (eds), Quantified Land Evaluation Procedures. ITC Publication, No. 6, 29–35.

Molina, J. A. A., Clapp, C. E., Shaffer, M. J., Chichester, F. W. and Larson, W. E. (1983) NISOIL: A model of nitrogen and carbon transformations in soil: Description, calibration and behaviour. Soil Science Society America Journal, 47, 85–99.

Nix, J. (1990) Farm management pocketbook. 20th Edition Farm Business Unit, Department of Agricultural Economics, Wye College, University of London.

Parton, W. J., Schimel, D. S., Cole, C. V. and Ojima, D. S. (1987) Analysis of factors controlling soil organic matter levels in the Great Plains grasslands. Soil Science Society America Journal, 51, 1173–1179.

Sanchez, P. A. and Miller, R. H. (1986) Organic matter and soil fertility management in acid soils of the tropics. Transactions of XIII Congress of Soil Science, Hamburg. Volume VI, 609–625.

Thomasson, A. J. and Jones, R. J. A. (1989) Computer mapping of soil trafficability in the UK. In: Jones, R. J. A. and Biagi, B. (eds.), Proceedings of EC Symposium on Computerization of Land Use Data, Pisa, Italy, 1987.

Veroney, R. P., van Veen, J. A. and Paul, E. A. (1981) Organic carbon dynamics in grassland soils. 2. Model validation and simulation of the long term effects of cultivation and rainfall erosion. Canadian Journal of Soil Science, 61, 211–224.

Whittle, I.R. (1990) Lands at risk from sea level rise in the UK. In: Doornkamp, J.C. (ed.), The Greenhouse Effect and Rising Sea Levels in the United Kingdom. M1 Press, Long Eaton, Notts., 85–93.

*Section 4*

Cannell, M. G. R., Grace, J. and Booth, A. (1989) Possible impacts of climatic warming on trees and forests in the United Kingdom: a review. Forestry, 62, 337–364.

Cannell, M. G. R. and Hooper, M. D. (1990) The greenhouse effect and terrestrial ecosystems in the UK. Institute of Terrestrial Ecology Research Publ. No. 4, HMSO, London.

Eamus, D. and Jarvis, P. G. (1989) The direct effects of increase in global atmospheric $CO_2$ concentration on natural and commercial temperate trees and forests. Advances in Ecological Research, 19, 2–47.

Ford, M. J. (1978) A study of the biological response to climatic change. Nature Conservancy Council, Peterborough.

George, D. G. (1988) Impacts of climatic change on freshwater ecosystems. Freshwater Biological Association, Ambleside, Cumbria.

Grime, J. P. and Callaghan, T. V. (1988) Direct and indirect effects of climatic change on species ecosystems and processes of conservation and amenity interest. Contract Report to the Department of the Environment, London.

Grime, J. P., Hodgson, J. C. and Hunt, R. (1988) Comparative plant ecology: a functional approach to common British species. Unwin Hyman, London.

Hodgson, J. G. (1986a) Commonness and rarity in plants with special reference to the Sheffield flora. I. The identity, distribution and habitat characteristics of the common and rare species., Biological Conservation, 36, 199–252.

Hodgson, J. G. (1986b) Commonness and rarity in plants with special reference to the Sheffield flora. II. The relative importance of climate, soils and land use. Biological Conservation, 36, 253–274.

Jarvis, P. G. (1989) Atmospheric carbon dioxide and forests. Phil. Trans. R. Soc. Lond. B324, 369–392.

Jeffree, E. P. (1960) Some long-term means from the phenological reports (1891–1948) of the Royal Meteorological Society. Q. J. R. Met. Soc., 86, 95–103.

Margary, I. D. (1926) The Marsham phenological record in Norfolk, 1736–1925, and some others. Q.J.R. Met. Soc., 52, 27–54.

Matthews, J. R. (1955) Origin and distribution of the British flora. Hutchinson, London.

Parry, M. L., Carter, T. R. and Porter, J. H. (1989) The greenhouse effect and the future of UK agriculture. J. Royal Agric. Soc. England, 120–131.

Perring, F. H. and Walters, S. M. (1962) Atlas of the British flora. Nelson, Edinburgh.

Rodwell, J. S. (1990) British plant communities (5 volumes). Cambridge University Press, Cambridge.

Southwood, T. R. E. (1962) Migration of terrestrial arthropods in relation to habitat. Biol. Rev., 37, 171–214.

Woodward, F. I. (1987) Climate and plant distribution. Cambridge University Press, Cambridge.

*Section 5*

Cannell, M. G. R., Grace, J. and Booth, A. (1989) Possible impacts of climatic warming on trees and forests in the United Kingdom: a review. Forestry, 62, 337–364.

Day, W. and Atkin, R. K. (eds.) (1985) Wheat growth and modelling. NATO ASI Series A, Volume 86, Plenum Publishing, New York.

Goudriaan, J. and Unsworth, M. H. (1990) Implications of increasing carbon dioxide and climate change for agricultural productivity and water resources. In: Impact of Carbon Dioxide, Trace Gases and Climate Change on Global Agriculture, ASA Special Publication no. 53, 111–130.

Harrington, R. (1989) Greenhouse greenfly? Antenna, Royal Entomological Society, London, 13, 169–172.

Kimball, B. A., Mauney, J. R., Radin, J. W., Nakayama, F. S., Idso, S. B., Hendrix, D. L., Akey, D. H., Allen, S. G., Anderson, M. G. and Harting, W. (1986) Effects of increasing atmospheric $CO_2$ on the growth, water relations, and physiology of plants grown under optimal and limiting levels of water and nitrogen. In: Response of Vegetation to Carbon Dioxide. Ser. 039. US Dept. of Energy and USDA-ARS, Washington, DC.

Long, S. P. (1990) Personal communication.

Mitchell, R. (1989) The effects of climate change on wheat yields: sensitivity of model predictions to increases in cloud cover. In: Bennett, R. M. (ed.), The Greenhouse Effect and UK Agriculture, Centre for Agricultural Strategy, University of Reading, Reading.

Morison, J. L. (1988) Effect of increasing atmospheric $CO_2$ on plants and their responses to other pollutants, climatic and soil factors. Aspects of Applied Biology, 17, 113–122.

Monteith, J. L. (1981) Climatic variation and the growth of crops. Quart. J. R. Met. Soc., 107, 749–774.

Monteith, J. L. and Scott, R. K. (1982) Weather and yield variation of crops. In: Blaxter, K. (ed.), Food, Nutrition and Climate. Elsevier Applied Science, England.

Parry, M. L., Carter, T. R. and Porter, J. H. (1989) The greenhouse effect and the future of UK agriculture. J. Royal Agric. Soc. England, 120–131.

Porter, J. (1984) A model of canopy development in winter wheat. J. Agric. Sci., Cambridge, 102, 383–392.

Porter, J. (1989) Interactions between climate change and nitrate leaching using the AFRC model of winter wheat growth. In: Bennett, R. M. (ed.), The Greenhouse Effect and UK Agriculture, Centre for Agricultural Strategy, University of Reading, Reading.

Squire, G. R. and Unsworth, M. H. (1988) Effects of $CO_2$ and climate change on agriculture. Report to UK Department of the Environment, University of Nottingham, School of Agriculture, Loughborough.

Unsworth, M. H., Cox, J. S. and Scott, R. K. (1989) Impacts on agriculture and horticulture. In: Impacts of the Mild Winter 1988/89. Report to the UK Department of the Environment, Institute of Terrestrial Ecology, Bush Estate, Penicuik, Scotland, 63–88.

Whittle, I. R. (1990) Lands at risk from sea level rise in the UK. In: Doornkamp, J. C. (ed.), The Greenhouse Effect and Rising Sea Level in the UK. M1 Press, Long Eaton, UK, 85–94.

*Section 6*

Beran, H. A. and Arnell, N. W. (1989) Effect of climatic change on quantitative aspects of United Kingdom water resources. Report to the Department of the Environment. Institute of Hydrology, Wallingford, Oxon.

Boorman, L. A., Goss-Custard, J. D. and McGrorty, S. (1989) Climatic change, rising sea level and the British coast. Institute of Terrestrial Ecology Research Publ. No. 1., HMSO, London.

Department of the Environment (1988) Possible impacts of climate change on the natural environment in the United Kingdom, London.

Doornkamp, J. C. (ed.) (1990) The greenhouse effect and rising sea levels in the United Kingdom. M1 Press, Long Eaton, Notts.

Edmunds, M. (1989) Personal communication.

Gilbert, S. and Horner R. (1984) The Thames Barrier. Thomas Telford, London.

Graff, J. (1981) An investigation of the frequency distribution of annual sea level maxima at ports around Great Britain. Estuarine, Coastal and Shelf Science, 12, 389–449.

National Research Council, (1987) Responding to Changes in Sea Level, Engineering Implications. National Academic Press, Washington DC.

Price, M. and Reed, D. W. (1989) The influence of mains leakage and urban drainage on groundwater levels beneath conurbations in the United Kingdom. Proceedings of the Institution of Civil Engineers, Part 1, 86, 31–39.

Pugh, D. T. (1987a) Tides, Surges and Mean Sea Level: a handbook for engineers and scientists. John Wiley & Sons, Chichester.

Pugh, D. T. (1987b) The global sea level observing system. Hydrographic Journal, 45, 5–8.

Pugh, D. T. (1990) Is there a sea level problem? Proceedings of the Institution of Civil Engineers, Part 1, 88, 347–366.

Stewart, R. W., Kjerfue, B., Millimam, J. and Dwivedi, S. N. (1990) Relative sea level change: a critical evaluation. UNESCO Reports in Marine Science, 54.

United States Department of Energy (1985) Glaciers, ice sheets and sea level: effect of a $CO_2$-induced climatic change. Report of a workshop held in Seattle, Washington, September 13–15, 1984. DOE/ER/60235-1.

Whittle, I. R. (1989) The greenhouse effect, land at risk, an assessment. Loughborough Conference of River and Coastal Engineers. Ministry of Agriculture, Fisheries and Food, London.

Wind, H. G. (ed.), (1987) Impact of Sea Level Rise on Society. A. A. Balkema, Rotterdam.

Woodworth, P. L. (1987) Trends in UK mean sea level. Marine Geodesy, 11, 57–87.

Woodworth, P. L. (1989) A Search for acceleration in records of European mean sea level. International Journal of Climatology, 10, 129–143.

*Section 7*

ACAH (1980) Water for agriculture and future needs. Advisory Council for Agriculture and Horticulture, London.

Arnell, N. and Reynard, N. (1989) Estimating the impacts of climatic change on river flows; some examples from Britain. Conference on Climate and Water, Helsinki, September, 1989.

Beran, M. A. (1982) Aspects of flood hydrology of the pumped fenland catchments of Britain. In: Proc. Symp. Polders of the World, Vol. 1, Lelystad, The Netherlands, p. 463.

Beran, M. and Arnell, N. (1989) Aspects of UK water resources. Report to the Department of the Environment, Institute of Hydrology, Wallingford.

BNA (1990) World Climate Change Report. Bureau of National Affairs, Paris.

CIRIA (1989) The engineering implications of rising groundwater levels in the deep aquifer beneath London. Construction Industry Research and Information Association, London.

CWPU (1976a) The Wash Barrage Storage Scheme: Report on the feasibility study. Central Water Planning Unit, Internal Report, Reading.

CWPU (1976b) The 1975–76 drought: a hydrological review. Central Water Planning Unit, Reading.

IWES/ICE (1977) Proceedings of the one-day seminar on the operational aspects of the drought of 1975–76. Institute of Water Engineers and Scientists/Institute of Civil Engineers, London.

Males, D. and Turton, P. (1979) Design flow criteria in sewers and water mains. Central Water Planning Unit, Reading.

NRA (1990) Annual Report, National Rivers Authority, London.

NWC (1977) The 1975–76 Drought, National Water Council, London.

Parry, M. L. and Read, N. J. (eds.) (1988) The impact of climatic variability on UK industry. AIR Report 1. Atmospheric Impacts Research Group, University of Birmingham, Birmingham.

WAA (1985) Drought '84. Water Authorities Association, London.

WRB (1973) Water Resources in England and Wales. Water Resources Board, Reading.

WSA (1989) Metering Trials Interim Report, No. 1, Water Services Association, London.

WSA (1990) Waterfacts. Water Services Association, London.

Wigley, T. M. L. and Jones, P. D. (1986) Recent changes in precipitation and precipitation variability in England and Wales, Report to the Water Research Centre by the Climatic Research Unit, University of East Anglia, Norwich.

*Section 8*

Chester, P. (1988) The potential for electricity from renewable energy sources. In: National Society for Clean Air, 55th Conference Proceedings, 24–27 October 1988, Llandudno, Wales.

Department of Energy (1983) Paper on energy projections methodology, Sizewell 'B' Public Inquiry, DEN/S/7 (NE).

Department of Energy (1989) An evaluation of energy-related greenhouse gas emissions and measures to ameliorate them. Energy Paper 58, London.

Edmonds, J. and Reilly, J. (1983) A long-term global energy-economic model of carbon dioxide release from fossil fuel use. Energy Economics, 5 (2), 74–88.

Electricity Council (1988) Annual Report and Accounts, 1987/88, Electricity Council, London.

Eunson, E. M. (1988) Proof of evidence on system considerations. Hinckley Point 'C' Power Station Public Inquiry, CEGB, September, 1988.

Grigg, P. (1990) Personal communication.

Grubb, M. (1989) A resource and siting analysis for wind energy in Britain. Proceedings of the European Wind Energy Conference, Glasgow.

Herring, H., Hardcastle, R. and Philipson, R. (1988) Energy use and efficiency in UK commercial and public buildings up to the year 2000. Energy Efficiency Series 6, Energy Efficiency Office, HMSO, London.

Milbank, N. (1989) Building design and use: response to climate change. Architects Journal, 96, 59–63.

IPCC Working Group II (1989) Likely impacts of climate change on human settlement; the energy, transport and industrial sectors: human health and air quality, and likely impacts of changes in UV-B. Report to IPCC Working Group 2 by Section 5.

Parry, M. L. and Read, N. J. (1988) The impact of climatic variability on UK industry, AIR Report 1, Atmospheric Impacts Research Group, University of Birmingham.

*Section 9*

Anon. (1990) Quarry Management, April, 1990.

BACMI (1989) Statistical Yearbook 1989. British Aggregate Construction Materials Industry, London.

British Coal Corporation (1989) Reports and Accounts 1988/89, British Coal Corporation, London.

British Geological Survey (1989) United Kingdom Minerals Yearbook. Kegworth, Nottingham.

Taylor, B. G. S. (1990) Personal communication.

Whittaker, B. N. (1990) UK coal challenge for the 90's: resources, reserves and markets. University of Wales Seminar, 2 March, 1990.

*Section 10*

Central Statistical Office (1989) UK National Accounts, HMSO, London.

Confederation of British Industry (1989a) Managing the Greenhouse Effect. CBI, London.

Confederation of British Industry (1989b) The Greenhouse Effect and Energy Efficiency. CBI, London.

Department of Energy (1989a) Digest of Energy Statistics, HMSO, London.

Department of Energy (1989b) Industrial Energy Markets, HMSO, London.

Department of the Environment (1989) Digest of Environmental Protection and Water Statistics, HMSO, London.

Palutikof, J. (1983) Impact of weather and climate on industrial production in Great Britain, Jnl. of Climatology, Vol. 3, 65–79.

Water Services Association (1989) Waterfacts. Water Services Association, London.

*Section 11*

Boden, J. B. and Driscoll, R. M. C. (1987) House foundations – a review of the effect of clay soil volume on design and performance. Municipal Engineering, 4, 181–213.

BSI (1988) BS 6399: Loading of Buildings, Part 3: Code of practice for imposed roof loads. British Standards Institution, London.

BSI (1990) BS 6399: Loading for Buildings, Part 2: Wind loading. British Standards Institution, London.

Building Effects Review Group (1989) The effects of acid rain deposition on buildings and building materials in the United Kingdom. Report to the Department of the Environment, HMSO, London.

BRE (1988) Daylighting as a passive solar energy option; an assessment of its potential in non-domestic buildings. BRE Report No. 129/88. Building Research Establishment, Watford.

BRE (1989) The assessment of wind loads, Parts 1 to 7, BRE Digest No. 346, Building Research Establishment. HMSO, London.

BRE (1990) Climate and construction operations in the Plymouth area. BRE/CR 35/90. Building Research Establishment, Watford.

CIRIA (1989) The engineering implications of rising groundwater levels in the deep aquifer beneath London. Construction Industry Research and Information Association, London.

Cook, N. J. (1985) The designer's guide to wind loading of building structures. Part 1: Background, damage survey, wind data and structural classification. A BRE Report, Butterworth, London.

Fanchiotti, A. (ed.) (1990) European Reference Book on Daylighting. Commission of the European Communities, Directorate General XII for Science, Research and Development, Brussels.

Hammond, J. M. (1990) The strong winds experienced during late winter 1989/90 over the UK: historical perspectives. Meteorological Magazine, 119, 211–219.

Milbank, N. (1989) Building design and use: response to climatic change. Architects Journal, 96, 59–63.

Page, J. K. (ed.) (1990) Indoor environment: health aspects of air quality, thermal environment, light and noise. World Health Organisation, Geneva.

UK Review Group on Acid Rain (1987) Acid deposition in the United Kingdom, 1981–1985, Warren Spring Laboratory, Stevenage, Herts.

*Section 12*

Crawford, C. L. (1989) The effect of severe weather on the Scottish rail system. In: Smith, K. (ed.), Weather Sensitivity and Services in Scotland, Scottish Academic Press, Edinburgh.

Hopkins, S. S. and White, K. W. (1975) Extreme temperatures over the UK for design purposes. Met. Mag. 104, 94–104.

Palutikof, J. (1991) Weather and road accidents. In: Perry, A. H. and Symons, L., (eds.), Highway Meteorology. Spon Limited (in press).

Parker, K. T., Lord, W. B. H., Read, N. J. and Parsons, J. (1986) Social and economic responses to climatic variability in the UK. The Technical Change Centre, London.

Parry, M. L. and Read, N. J. (1988) The impact of climatic variability on UK industry. AIR report 1, Atmospheric Impacts Research Group, University of Birmingham, Birmingham.

Perry, A. H. (1981) Environmental Hazards in the British Isles. George Allen & Unwin, London.

Perry, A. H. and Edwards, J. (1988) Frequency of freeze-thaw cycles in recent winters in Great Britain. J. Meteorol., 13, 120–122.

Perry, A. H. and Symons, L. (1991) Highway Meteorology. Spon Limited (in press).

Smith, K. (1989) Weather Sensitivity and Services in Scotland. Scottish Academic Press, Edinburgh.

Smith, K. (1990) Weather sensitivity of rail transport, In: World Meteorological Organisation, Economic and Social Benefits of Meteorological and Hydrological Services. WMO, No. 733, Geneva.

*Section 13*

Arnell, N. W., Clark, M. J., and Gurnell, A. M. (1984) Flood insurance and extreme events: the role of crisis in prompting changes in British institutional response to flood hazard. Appl. Geography, 4, 167–181.

Buller, P. S. J. (1988) The October gale of 1987: damage to buildings and structures in the south east of England. Building Research Establishment, Department of the Environment, London.

Davies, R. (1983) Flooding in Britain: review of causes, damage and control measures. The Insurance Technical Bureau, September, 1983, ITB R83/106.

Dimmock, G. (1989) The Reinsurer's view. In: The Greenhouse Effect – Implications for Insurers. The Insurance and Reinsurance Research Group, London.

Dlugolecki, A. F. (1989) The Insurer's view. In: The Greenhouse Effect – Implications for Insurers. The Insurance and Reinsurance Research Group, London.

Hempsell, M. (1961) The assessment of flood risk in the UK. J. Chartered Inst. Insurance, March, 1961, 115–137.

Hitchcock, S. (1989) Lecture to the Insurance Institute of London, 9th March, 1989 (unpublished, available from the Insurance Institute of London).

Roe, G. M. (1990) Report on the Breach to the Sea Wall in Towyn, North Wales on 26 February, 1990. Paper given at the Ministry of Agriculture Food and Fisheries Conference of River and Coastal Enquiries in 1989/90 Winter Floods, July 1990, Loughborough.

Shearn, W. G. (1990) Changing weather patterns – what can insurers do? In: Changing Weather Patterns – What Does This Mean For the Insurance Industry. The Insurance and Reinsurance Research Group (in press), London.

*Section 14*

Boorman, L. A., Goss-Custard, J. D. and McGrorty, S. (1989) Climate change, rising sea level and the British coast. Institute of Terrestrial Ecology Research Publ. No. 1, HMSO, London.

de Freitas, C. R. (1990) Recreation climate assessment. International Journal of Climatology, 10, 89–103.

Hay, B. (1989) Tourism and the Scottish weather. In: Harrison, S. J. and Smith, K., (eds.), Weather Sensitivity and Services in Scotland. Scottish Academic Press, Edinburgh, 162–166.

Leatherman, S. P. (1989) Beach response strategies to accelerated sea-level rise. In: Topping, J. C. (ed.), Coping with Climate Change. Climate Institute, Washington, DC.

Masterton, J. M. and McNichol, D. W. (1981) A recreation climatology of the national capital region. Climatological Studies No. 34, Atmospheric Environment Service, Environment Canada, Toronto.

McBoyle, G. R., Wall, G., Harrison, R., Kinnaird, V. and Quinlan, C. (1986) Recreation and climate change: a Canadian case study. Ontario Geography, 28, 51–68.

Paul, A. H. (1972) Weather and daily use of outdoor recreation areas in Canada. In: Taylor, J. A. (ed.) Weather Forecasting for Agriculture and Industry. David and Charles, Newton Abbot.

Smith, K. (1988) Avalanche hazards: The rising death toll. Geography, 73, 157–158.

Smith, K. (1990) Tourism and climate change, Land Use Policy, 7(2), 176–180.

# Glossary

This glossary covers terms specific to climate change and sea level rise and their associated impacts.

**Adsorption**
Adherence of atoms, ions or molecules of a liquid or a gas to the surface of another substance (the adsorbent).

**Algal Bloom**
Proliferation of algae (seaweeds etc.) which remove oxygen from bodies of water, leaving the latter unable to support many life forms.

**Aquifer**
A layer of rock which holds water and allows water to percolate through it.

**Carbon Cycle**
The circulation of carbon atoms in nature, brought about mainly by the physical and chemical (metabolic) processes of living organisms.

**Degree Days**
The number of degrees below (or above) a threshold temperature, accumulated over all the days in the year (or during a given period) on which the mean temperature is below (or above) the threshold value.

**Determinate (crops)**
Crops in which each successive growth stage is finite.

**Dry Deposition**
The direct transfer of particulate and gaseous material from the atmosphere to a surface, without indirect input in aqueous solution or suspension.

**Evapotranspiration**
Evaporation of water vapour from a plant surface.

**Feedback**
The return to the input of a process of part of the output.

**Field Capacity**
A state of the soil in which all pores are recharged with water and there is no soil moisture deficit.

**Indeterminate (crops)**
Crops where growth stages are not finite and continue as long as conditions permit.

**Isotherm**
A line on a map joining points having the same temperature at the same point in time (or the same mean temperature over a given period of time).

**Leaching**
The loss of soluble substances from matter (e.g. soil) by the passage through of liquid (e.g. rainwater).

**Littoral**
Strip of land along sea coast lying between the extreme low and high tide marks, or the zone of a lake or pond from the water's edge down to about 6 metres.

**Mineralization (soils)**
The oxidation of organic compounds (i.e. carbon-containing compounds) in the soil to produce inorganic ions.

**pH**
A measure of the concentration of hydrogen ions in a solution. The pH of pure water is 7 (neutrality). A pH below 7 indicates acidity, of greater than 7 indicates alkalinity.

**Phenology**
The physical attributes of an organism; a classification based on observed similarities and differences in appearance, as opposed to genetic (evolutionary) similarities.

**Photorespiration**
A light-dependent respiration, the chemistry of which differs from normal (dark) respiration, that occurs in most photosynthetic plants. It results in the loss of $CO_2$ and energy which in turn reduces photosynthetic efficiency.

**Photosynthesis**
The synthesis of organic compounds by the reduction of carbon dioxide (i.e. the addition of hydrogen to $CO_2$), using light energy absorbed by chlorophyll.

**Stomata**
Pores or apertures in the outer cell layer of the aerial parts of leaves, stems and flowers through which water is lost by transpiration and gas exchange takes place.

**Troposphere**
The lower layers of the atmosphere extending from the earth's surface to a height of about 10 km.

**Ultraviolet (UV)**
Electromagnetic radiation having a wavelength between the violet end of the visible spectrum and X-rays ($\sim 10^{-7} - 10^{-9}$ m).

**Wave refraction**
Changes in wave direction as a result of changing water depth.

# Acronyms

| | |
|---|---|
| AFRC | Agricultural and Food Research Council |
| BP | Before Present |
| BRE | Building Research Establishment |
| BSI | British Standards Institution |
| CBI | Confederation of British Industry |
| CFCs | Chlorofluorocarbons |
| $CH_4$ | Methane |
| CIRIA | Construction Industry Research and Information Association |
| $CO_2$ | Carbon Dioxide |
| DNA | Deoxyribonucleic Acid |
| ESRC | Economic and Social Research Council |
| GCM | General Circulation Model |
| GLOSS | Global Sea-Level Network |
| GtC | Gigatonnes of Carbon |
| GW | Gigawatt |
| ICE | Institute of Civil Engineers |
| IPCC | Intergovernmental Panel on Climate Change |
| ISO | International Standards Organisation |
| ITE | Institute of Terrestrial Ecology |
| IWES | Institute of Water Engineers and Scientists |
| MVA | Megavoltamperes |
| NCC | Nature Conservancy Council |
| NERC | Natural Environment Research Council |
| NIPS | National Ice Prediction System |
| NNR | National Nature Reserve |
| $NO_x$ | Nitrogen Oxides |
| NRA | National Rivers Authority |
| ODN | Ordnance Datum Newlyn |
| PJ | Petajoules |
| ppmv | Parts per million by volume |
| PSMD | Potential Soil Moisture Deficit |
| SERC | Science and Engineering Research Council |
| $SO_2$ | Sulphur Dioxide |
| $SO_4$ | Sulphate |
| SSSI | Site of Special Scientific Interest |
| UNEP | United Nations Environment Programme |
| WMO | World Meteorological Organisation |

# Membership of Climate Change Impacts Review Group

Professor M. L. Parry *(Chairman)*
Atmospheric Impacts Research Group
School of Geography
University of Birmingham
Birmingham B15 2TT

Professor P. Bullock
Soil Survey and Land Research Centre
Silsoe
Bedfordshire MK45 4DT

Dr. M. G. R. Cannell
Institute of Terrestrial Ecology
Bush Estate
Penicuik
Midlothian ED26 0QB

Dr. A. F. Dlugolecki
General Accident Fire and Life Assurance Corporation, plc
Pithealvis
Perth PH2 0NH

Dr. G. Marshall
Dobson Park Industries plc
Dobson Park House
Manchester Road
Ince
Wigan WN2 2DX

Professor J. Page
The Cambridge Interdisciplinary Environmental Centre
Department of Geography
University of Cambridge
Downing Site
Cambridge CB2 3EA

Dr. A. Perry
Department of Geography
University College Swansea
Swansea SA2 6TT

Professor D. Potts
Department of Mining Engineering
University of Nottingham
Nottingham NG7 2RD

Dr. D. Pugh
Deacon Laboratory
Institute of Oceanographic Science
Brook Road
Wormley
Surrey GU8 5UB

Dr. J. F. Skea
Science Policy Research Unit
Mantell Building
University of Sussex
Falmer
Brighton BN1 9RF

Professor K. Smith
Department of Environmental Science
University of Stirling
Stirling FK9 4LA

Mr. P. Turton
Water Services Association
1 Queen Anne's Gate
London SW1H 9BT

Professor M. H. Unsworth
Department of Physiology and Environmental Science
School of Agriculture
University of Nottingham
Sutton Bonnington
Loughborough LE12 5RD

Dr. R. Warrick
Climatic Research Unit
University of East Anglia
Norwich NR4 7TG

Dr. S. M. Cayless *(Technical Secretary)*
Department of the Environment
Romney House
43 Marsham Street
London SW1P 3PY

Dr. R. B. Wilson *(Executive Secretary)*
Department of the Environment
Romney House
43 Marsham Street
London SW1P 3PY

# List of persons consulted

**SECTION 2 (Future Changes in Climate and Sea Level) Lead Author: R. Warrick.**
Persons consulted: E. Barrow; M. Hulme; P. D. Jones; S. C. B. Raper; T. M. L. Wigley.

**SECTION 3 (Soils) Lead Author: P. Bullock.**
Persons consulted: E. M. Bridges; D. Dent; D. J. Greenwood; O. W. Heal; P. Ineson; D. Jenkins; D. Powlson; D. L. Rimmer; G. Spoor; R. S. Swift; T. R. E. Thompson; M. J. Wilson.

**SECTION 4 (Flora, Fauna and Landscape) Lead Author: M. G. R. Cannell.**
Persons consulted: M. Beran; K. Bull; D. A. Burdekin; T. V. Callaghan; A. Cooke; I. Cooper; C. Cummins; P. H. Freer-Smith; A. Gray; J. P. Grime; P. Harding; O. W. Heal; G. A. F. Hendry; M. Hill; J. G. Hodgson; R. Hunt; P. G. Jarvis; D. Jeffries; H. Kruuk; T. A. Mansfield; T. Mitchell Jones; M. Morris; M. Neugent; I. Newton; T. M. Roberts; A. Stubbs; A. D. Watt; S. J. Woodin; F. I. Woodward.

**SECTION 5 (Agriculture, Horticulture, Aquaculture and Forestry) Lead Author: M. H. Unsworth.**
Persons consulted: M. Goss; T. Lewis; B. Marshall; J. Morison; J. Porter; R. J. Roberts; G. J. A. Ryle; R. K. Scott; C. J. Wright.

**SECTION 6 (Coastal Regions) Lead Author: D. Pugh.**
Persons consulted: A. Allison; M. G. Barrett; A. Bullock; L. Draper; M. Edmunds; A. Hawkins; J. J. A. Jones; F. Law; I. Shennan; J. Shepherd; I. Townend; P. Woodworth.

**SECTION 7 (Water Industry) Lead Author: P. Turton.**
Persons consulted: N. Arnell; A. H. Bunch; M. Carney; D. J. Cook; J. M. Davis; S. M. Postle; J. Sherriff; P. Stott.

**SECTION 8 (Energy) Lead Author: J. F. Skea.**
Persons consulted: P. Bartlett; S. Boyle; N. A. Burdett; I. Fells; P. Grigg; M. J. Grubb; G. Henderson; H. Herring; J. Hirst; P. M. S. Jones; P. O. Lewis; M. J. Parker; C. Swadkin.

**SECTION 9 (Minerals Extraction) Lead Author: D. Potts.**
Persons consulted: T. S. Brewer; J. S. Edwards; I. Longson; G. E. Pearse; B. Taylor; P. J. Cook; J. F. Tunnicliffe.

**SECTION 10 (Manufacturing) Lead Author: G. Marshall.**
Persons consulted: K. Alston; G. Barrett; N. Burdett; D. Cook; J. Frame; M. J. S. Gibson; D. Green; K. Gregory; P. O. Lewis; J. Marsh; A. Needham; N. Read; S. Pullen; P. Rowlatt; A. J. Warren.

**SECTION 11 (Construction) Lead Author: J. Page.**
Persons consulted: P. Bartlett; D. Bland; K. W. G. Blount; P. L. Bransby; W. Eastwood; F. Garas; T. Hanna; D. Hawkes; J. Lawson; D. MacGowan; N. Milbank; J. Miller; L. Oliver; L. Olsen; M. J. Prior; D. Robson; M. Roy; A. J. Saul; R. Taeslar; S. H. Tasker; V. B. Torrance.

**SECTION 12 (Transport) Lead Author: A. Perry.**
Persons consulted: J. C. Field; A. C. Howard; R. D. Hunt; A. W. Lock; R. Neil; R. J. Ward.

**SECTION 13 (Financial Sector) Lead Author: A. F. Dlugolecki.**
Persons consulted: D. A. Dennett; G. Dimmock; M. Fearnsides; D. Foreman; J. C. Lodge; M. W. Oakes; A. G. C. Paish; H. Thompson.

**SECTION 14 (Recreation and Tourism) Lead Author: K. Smith.**
Persons consulted: J. G. L. Adams; A. Brazewell; E. N. Brogan; C. R. de Freitas; S. Lewis; N. Poole; J. Ward.

# Reports prepared at the request of the Department of the Environment

**Acid Deposition in the United Kingdom**
UK Review Group on Acid Rain, December 1983

**Acid Deposition in the United Kingdom 1981–1985**
UK Review Group on Acid Rain, August 1987

Copies of both available from
Warren Spring Laboratory
Gunnels Wood Road
Stevenage
Herts
SG1 2BX
Price £10.00 each.

**Acidity in United Kingdom Fresh Waters**
UK Acid Waters Review Group, April 1986.

**Ozone in the United Kingdom**
UK Photochemical Oxidants Review Group, February 1987.

**Acid Deposition in the United Kingdom 1986–1988**
UK Review Group on Acid Rain, September 1990

**Oxides of Nitrogen in the United Kingdom**
Second Report of the UK Photochemical Oxidants Review Group, June 1990

Copies available from
DOE/DTP Publications Sales Unit
Building 1
Victoria Road
South Ruislip
Middlesex
HA4 0NZ
Price £10.00 each.

**Stratospheric Ozone**
UK Stratospheric Ozone Review Group, August 1987.

Copies available from HMSO (ISBN 0 11 752018 7)
Price £9.95 each

**Stratospheric Ozone 1990**
UK Stratospheric Ozone Review Group, June 1990

Copies available from HMSO (ISBN 0 11 752320 8)
Price £5.30 each

**The Effects of Acid Deposition on Buildings and Building Materials in the UK**
Building Effects Review Group.

Copies available from HMSO (ISBN 0 11 752178 7)
Price £11.25 each

**The Effects of Acid Deposition on the Terrestrial Environment in the United Kingdom.**
UK Terrestrial Effects Review Group, 1988.

Copies available from HMSO (ISBN 0 11 752029 2)
Price £11.95 each

Printed in the United Kingdom for HMSO.
Dd.0293024, 1/91, C50, 3385/4, 5673, 136547.